THE
KIMCHI
BOOK

김치 책

고은정

장안에 내로라하는 음식 전문가들도 김치와 장 담 그기를 배우러 지리산 인근, 고은정의 '맛있는 부 엌'을 찾는다. 저자가 가르치는 제철음식학교, 시의 적절약선학교, 우리장학교, 김치학교 등의 수업은 늘 일찌감치 마감되어 등록이 어려울 지경이다.

저자는 제철 식재료에 대한 이해를 통해 제대로 된 밥 짓기와 제철 김치 담그기, 직접 담근 장으로 조 리하기 등을 30여 년째 교육해 왔으며 2017년부터 서울시 '장하다 내 인생'을 통해 공동 장독대 사업 을 해왔다. 이후 청와대가 개방되기 전까지 청와대 장독대를 복원하고 청와대 조리사들에게 장 담그 는 방법을 교육했다.

저자가 말하는 김치는 특별한 비법으로 완성되는 음식이 아니다. 제철에 난 재료로 '라면 끓이듯' 쉽 게 담그며, 발효가 스스로 길을 찾도록 지켜보는 일 에 가깝다. 『김치 책』은 그가 부엌과 장독대에서 반 복해 온 이 기본적인 태도를 정리한 기록이다.

Goh Eun-jeong

Even the most renowned food experts in the country travel to Goh Eun-jeong's "Delicious Kitchen," located near Jirisan mountain, to learn the art of making kimchi and *jang* (fermented soybean paste and sauce). Her classes — including the Seasonal Food School, Seasonal Medicinal Food School, Our Jang School, and Kimchi School — fill up so quickly that securing a spot is widely known to be a challenge.

For over thirty years, Goh has taught students how to cook rice properly, make seasonal kimchi, and cook with homemade *jang*, all rooted in a deep understanding of seasonal ingredients. Since 2017, she has led a community *jangdokdae* (fermentation jar platform) project as part of the Seoul Metropolitan Government's "My Great Life!" campaign. She also restored the *jangdokdae* at the Blue House (Cheong Wa Dae) and trained the presidential chefs in traditional *jang* making before the residence was opened to the public.

To Goh, kimchi is not a dish perfected by secret recipes. Rather, she believes it should be as easy to prepare as ramen instant noodles, using ingredients available in season. It is less about cooking and more about watching and waiting as fermentation finds its own path. *The Kimchi Book* is a record of this fundamental philosophy, honed over years in her kitchen and among her fermentation jars.

일러두기

찹쌀 풀 찹쌀 풀 1컵을 기준으로 할 때 냄비에 찹쌀가루 1 큰술과 물 1컵을 넣고 잘 푼 뒤 불에 올리고 저으면서 풀을 쑤어 식혀 사용한다. 멥쌀 풀은 멥쌀가루, 밀가루 풀은 밀가 루를 사용해 찹쌀 풀과 같은 방법으로 쑤면 된다.

배즙 이 책에서는 시판 배즙(갈색) 또는 배 음료(투명한 색) 를 사용했다.

생강술 껍질을 벗긴 생강을 곱게 찧어 청주에 넣고 반나절 이상 재웠다가 체에 밭쳐 건더기를 걸러내고 액체만 사용한 다(생강술의 비율은 청주 3~4 : 생강 1).

육수 물 1L, 멸치 5마리, 다시마 1조각(10cm×10cm), 표 고버섯 1~2개, 자투리 채소 약간을 냄비에 넣고 센불로 끓 이다가 끓기 시작하면 약불로 바꿔 15분간 더 끓인 후 체에 국물만 걸러내 냉장 보관해 두고 쓴다.

Notes

Glutinous rice paste: For 1 cup of glutinous rice paste: whisk 1 tablespoon of glutinous rice powder into 1 cup of water in a pot, boil and stir until the mixture thickens into a paste. Remove from the heat and cool before use. Use the same proportions and method with non-glutinous rice powder to make regular rice paste, or with flour to make a flour paste.

Pear juice: In this book, commercially available pear juice (brownish color) or a clear pear beverage was used.

Ginger liquor: Peel and mince fresh ginger, then steep it in rice wine for at least half a day. Strain the mixture through a fine sieve and discard the solids; use only the infused liquor. Ratio: 3–4 parts rice wine to 1 part ginger.

Broth: Combine 1 L of water, 5 dried anchovies, one piece of kelp (about 10 cm×10 cm), 1–2 dried shiitake mushrooms, and a handful of vegetable trimmings in a pot. Bring to a boil over high heat, then reduce to low and simmer for 15 minutes. Strain the broth, cool it, and store it in the refrigerator until needed.

THE KIMCHI BOOK

김치 책

Mastering
The Basics of
Kimchi with
The Chefs'
Teacher

셰프들의
김치 선생님,
고은정의
기본 김치 레시피

몽스북
mons

여름은 모질고 겨울은 매섭습니다. 산이 7할을 차지해 부쳐 먹을 땅도 넉넉지 않은 데다, 토양에 유기물이 적어 거름을 부지런히 주지 않으면 만족할 만한 소출을 기대하기 어렵습니다. 한반도 사람들이 오랫동안 다양한 푸성귀를 먹어온 이유입니다. 푸성귀를 밥과 어우러지게 하고, 오래 보관하는 데에는 김치만 한 음식이 없습니다. 김치를 제대로 이해하려면, 그래서 한반도의 자연을 먼저 알아야 합니다. 고은정 선생이 지리산 자락에 자리를 잡은 것은 한반도의 자연을 자신의 몸에 들이겠다는 뜻임을 저는 잘 알고 있습니다. 그의 김치에는 한반도의 자연이 담겨 있습니다. 한반도에서의 삶을 이해하고자 한다면 그의 김치를 맛보아야 합니다.

— 황교익(맛칼럼니스트)

여전히 김치 레시피에는 손맛과 눈대중이 통한다. 내림이라 그렇고, 전통이라는 신비한 세계에 속해 있어 그렇다. 다 좋은데, 그 때문에 쉽게 만들어보지 못한다. 어찌어찌 따라 해도 실패다. 운 좋게 고은정 선생의 김치를 몇 번 먹어봤다. 허다한 김치 가운데, 아플 때 생각나는 김치가 그의 김치다. 왜 그런지는 모르겠다. 한 번만 더 먹어봤으면! 그 소망이 이루어지려나. 딱 부러지는 계량을 기준으로 한 고은정 선생의 『김치책』이 나왔다. 시키는 대로만 하면 된단다. 야호!

— 박찬일(셰프, 작가)

김치 담그기가 엄두 나지 않고 자신 없을 때, 그런데 맛있는 김치는 먹고 싶고 엄마 김치가 그리울 때! 쉽고 맛있게 담글 수 있다고 알려주는 간결한 레시피에 도전 정신까지 불러일으키는 『김치 책』. 이 책대로 하면 누구나 산과 들과 바다의 재료로 맛있는 김치를 담글 수 있을 것이다. 너무 맛있어서 깜짝 놀랄지도 모른다. 겉절이부터 김장 김치까지, 당장 도전!

— 양희경(배우)

Recommendations

The summers are harsh, and the winters are severe. With mountains covering seventy percent of the land, arable soil is scarce and poor in organic matter; without diligent fertilization, a bountiful harvest is hard to come by. This is why the people of the Korean Peninsula have long relied on such a diverse array of wild greens. And when it comes to harmonizing these greens with rice and preserving them for the long term, nothing compares to kimchi..

Therefore, to make good kimchi, one must truly understand the nature of the Korean Peninsula. I know well that Goh Eun-jeong settled at the foot of Mount Jiri to let the nature of this land permeate her very being. Her kimchi captures the essence of the Korean landscape.

To understand life on the Korean Peninsula, you must taste her kimchi.

— **Hwang Kyo-ik (Food columnist)**

Kimchi recipes still rely heavily on sonmat (handmade flavor) and rough estimation. This is because they are a legacy passed down, part of a world of traditional mystery. That is all well and good, but it is also the reason most people never dare to make it themselves. Even if they try, it often ends in failure.

I have been lucky enough to taste Goh Eun-jeong's kimchi a few times. Of all the countless kinds of kimchi out there, hers is the one I crave when I am sick. I don't know exactly why. I just know I always found myself wishing, 'If only I could have that one more time!'

Now, that wish may finally come true. This is a kimchi book built on precise, clear-cut measurements. They say all you have to do is follow the instructions — and it works!

— **Park Chan-il (Chef, Author)**

For when the thought of making kimchi feels overwhelming and you lack confidence, yet you still crave that delicious flavor and miss your mom's kimchi! This book offers concise recipes that show you just how easy and tasty it can be, inspiring you to take on the challenge. You will soon find yourself making delicious kimchi with ingredients from the mountains, fields, and sea, and you'll be surprised by how incredible it tastes. From fresh geotjeori kimchi to aged *gimjang* kimchi, get started today!

— **Yang Hee-kyung (Actress)**

목차 | Contents

추천의 글	Recommendations	4
서문	Prologue	12
계절의 맛이 깃든 우리 김치 이야기	The Story of Kimchi Through the Seasons	16

봄 | Spring

고수김치	*Gosu*-Kimchi	28
참죽순김치	*Chamjuksun* Kimchi	34
미나리김치	*Minari* Kimchi	40
미나리물김치	*Minari Mul*-Kimchi	46
달래김치	*Dallae* Kimchi	52
좁쌀(기장)깍두기	*Jopssal (Gijang) Kkakdugi*	56
알배추물김치	*Albaechu Mul*-Kimchi	60
산갓물김치	*Sangat Mul*-Kimchi	66
봄동사과김치	*Bomdong Sagwa* Kimchi	70
상추겉절이	*Sangchu Geotjeori*	76

여름   Summer

오이소박이	*Oi-Sobagi*	82
가지소박이	*Gaji-Sobagi*	88
부추김치	*Buchu*-Kimchi	94
오이소박이물김치	*Oi-Sobagi Mul*-Kimchi	98
깻잎간장김치	*Kkaennip Ganjang* Kimchi	102
보리열무김치	*Bori Yeolmu*-Kimchi	108
열무물김치	*Yeolmu Mul*-Kimchi	114
얼갈이김치	*Eolgari*-Kimchi	120
얼갈이배추물김치	*Eolgaribaechu Mul*-Kimchi	126
토마토김치	Tomato Kimchi	132
고구마줄기김치	*Goguma Julgi* Kimchi	138
풋사과연근김치	*Putsagwa Yeongeun* Kimchi	144
오이지(오이지무침)	Oiji (*Oijimuchim*)	148
고추소박이	*Gochu Sobagi*	154
미니오이소박이	Mini *Oi-Sobagi*	160
양파김치	*Yangpa* Kimchi	166
양배추양파김치	*Yangbaechu Yangpa* Kimchi	170

가을   Autumn

마배깍두기	*Ma Bae Kkakdugi*	176
도라지제피김치	*Doraji Jepi* Kimchi	180
고들빼기김치	*Godeulppaegi*-Kimchi	186
배추김치	*Baechu*-Kimchi	192
백김치	*Baek*-Kimchi	202
늙은호박김치	*Neulgeun Hobak*-Kimchi	208
갓김치	*Gat*-Kimchi	214
단감김치	*Dangam* Kimchi	220
쪽파김치	*Jjokpa* Kimchi	226
총각김치	*Chonggak*-Kimchi	232
총각무물김치	*Chonggakmu Mul*-Kimchi	238

겨울　　　　Winter

굴깍두기　　　　*Gul–Kkakdugi*　　　　244

나박김치　　　　*Nabak*-Kimchi　　　　250

시금치겉절이　　　*Sigeumchi Geotjeori*　　256

동치미　　　　*Dongchimi*　　　　260

무생채　　　　*Musaengchae*　　　　266

배무생채　　　　*Bae Musaengchae*　　270

파래김치　　　　*Parae* Kimchi　　　274

서문

결혼을 하면서 나는 어머니로부터 '김치 독립'을 했다. 혼수로 사 간 370L짜리 냉장고에 의지해 식재료를 보관해 두고 먹으려니 절대적으로 용량이 부족해 일상 속 불편함이 있었다. 특히 김치는 적은 양을 자주 담가서 작은 통에 넣어 저장해야 했다. 직장을 쉬는 주말마다 김치를 담그는 일의 무한 반복이었다. 그래도 외가의 우물 속 김치를 생각하면 냉장고에 김치를 넣어두고 먹는 걸 감사하며 자주 부지런하게 김치를 담그는 일에 불평을 모르고 살았다.

그렇다고 김치를 잘 담그는 것은 아니었다. 매번 어머니가 김치 담그실 때 잔심부름을 하면서 어깨너머로 본 것들을 기억해 흉내 내는 수준이었다. 실패할 때도 있었고 너무 맛있게 담가져 김치통 속 김치가 줄어드는 것이 마냥 아쉬울 때도 있었다. 그런 실패와 성공이 모이고 쌓여 오늘의 내가 되었다고 생각한다.

돌이켜보면 그 모든 것의 바탕에 어머니가 계셨다. 어쩌다 친정에 가는 날이면 어느 하루도 빠지지 않고 어머니가 뚝딱 차려 주시는 밥상에 빠져 코를 박고 과식을 했던 기억이 전부였다. 나는 여러 시간을 허둥대며 음식을 해야 손님을 초대할 수 있었는데 어머니는 달랐다. 갑작스런 방문에도 불가사의할 만큼 언제나 밥상은 풍요로웠다. 특히 김치 몇 가지는 빠지지 않고 올랐다. 우리 집 냉장고에도 있는, 내가 담근 김치와는 차원이 다른 맛으로 다가왔다.

어머니는 내가 잘 먹던 김치를 놓치지 않고 보고 계셨다가 돌아가는 내 손에 그 김치를 챙겨 들려주셨다. 김치 독립을 했다고는 하나 어머니 김치에서 벗어나지 못하고 허우적거리며 지내왔다고 하는 것이 맞는 시간들이었다.

지리산의 품으로 이사를 하고 근처의 대안학교로 출근을 시작하게 되어 어머니께 근황을 말씀드렸던 날을 기억한다. 그러자 어머니는 거기서 뭘 가르치냐고 물으셨다. 살림학과를 맡게 되었다고 하자 망설이지 않고 툭 던지신 한 마디가 나에게 큰 상처를 냈고 오래도록 아물지 않았다.

"네까짓 게 살림에 대해 뭘 안다고!"

오래도록 어머니께 인정을 받고 싶었다. 어머니 말씀대로 살림에 대해서 뭣 하나 알지 못하는 내게 배울 학생들이 걱정되기도 했다. 그래서 잠을 설치고 준비를 하면서 하루하루 아슬아슬하게 보냈다. 그러다 십여 년 전 처음으로 어머니께 김장 김치를 보내드렸다. 팔십을 바라보는 어머니가 많은 시간과 힘을 필요로 하는 김장을 하시기 어려울 것 같아서다. 무작정 보내드린 김장 김치에 놀라시거나 맛에 대한 부정적인 반응이 올까 봐 걱정이 되었다. 그런데 의외로 "고맙다, 김치가 참 맛있구나." 하는 어머니의 인사가 전해졌다. 비로소 오랜 체증이 쑥 내려가는 순간이었다. 그리고 처음으로 내가 담그는 김치에 자부심을 가지게 되었다.

어머니께 인정받은 김치들과 계절의 맛이 담긴 김치들을 이 책에 담고 싶었다. 나처럼 오래도록 헤매지 않고도 사람들이 쉽게 김치를 담갔으면 하는 바람에서다. 그냥 라면 끓이듯이 김치를 담갔으면 좋겠다. 책이 나오면 가장 먼저 어머니께 보여드리고 다시 칭찬을 받고 싶다.

Prologue

Getting married marked the moment I became independent with my kimchi-making. The 370-liter refrigerator I brought as part of my dowry was never big enough to store kimchi, though, and it made my daily life rather inconvenient. I was constantly making small batches of kimchi and storing them in tiny containers. Every weekend, when I was off from work, I found myself in a never-ending cycle of making kimchi.

Yet, when I remembered the kimchi stored in the well at my maternal grandmother's house, I felt thankful just to have a refrigerator to store and eat kimchi from. So, I rarely complained about the endless work of making kimchi.

That doesn't mean I was particularly good at it. Each time, I tried to recall and recreate what I had seen over my mother's shoulder while helping her with little chores as she made kimchi. Sometimes I failed, and sometimes the kimchi turned out so delicious that I felt sad to see the kimchi disappearing from the container so quickly. I believe all those moments of failure and success have combined to shape the person I am today.

Looking back, I realized that everything started with my mother. Whenever I visited my parents' house, my clearest memories are of the meals my mother would prepare with such apparent ease. I would bury my face in the food and always end up overeating. While I would need hours of frantic preparation to host guests, my mother was different. Even with an unexpected visit, her table was always abundant. It felt almost magical — without fail, there were always several kinds of kimchi. Even when it was the same kind of kimchi I had in my own refrigerator, hers tasted like something out of this world. My mother always noticed which kimchi I enjoyed the most, and before I left, she'd pack it up for me to take home. Even though I claimed to be independent in terms of kimchi, in truth, I am still caught up in my mother's kimchi and have never truly broken free.

I remember the day I told my mother about our move to the foot of Jirisan Mountain and how I started working at a nearby alternative school. She asked what I would be teaching there. When I told her I would be in charge of the Home Economics department, she didn't hesitate to blurt out a comment that wounded me deeply, leaving a scar that took a long time to heal:

"What would someone like you know about homemaking?"

For a long time, I wanted to be acknowledged by my mother. I often worried about the students who had to learn from someone like me — someone who, as my mother had pointed out, knew so little about homemaking. So, I spent many restless nights preparing for my class, walking a tightrope each day. Then, about ten years ago, I sent my mother my own batch of *gimjang* kimchi for the first time. As she was approaching eighty, I thought it might be too physically demanding for her to go through the process of making *gimjang* kimchi herself. I worried she might be upset by this unexpected gesture or even criticize the taste. But to my surprise, her response was warm: "Thank you, the kimchi is really delicious." In that moment, it felt like a weight had finally been lifted from my shoulders. And for the first time, I felt proud of the kimchi I had made.

I wanted to include those kimchi recipes in this book — the ones my mother approved of, the ones that capture the distinct flavors of each season. It is my hope that people are able to make kimchi easily, without struggle — much like boiling a pot of ramen. When this book is published, I hope to show it to my mother first and receive her praise once again.

계절의 맛이 깃든 우리 김치 이야기

김치는 채소를 저장해 두고 먹기 위해 소금에 절여 가공하던 오래된 풍습에서 시작된 한국의 대표적인 전통 음식이다. 사계절의 변화가 뚜렷한 한반도의 기후 조건 속에서 채소를 오래 보존하기 위한 가공 기술이 발전한 것은 자연스러운 결과라고 할 수 있다. 우리나라의 여름은 기온과 습도가 높기 때문에 채소의 저장 기간을 늘려야 하는 절실함이 있었을 것이다. 또한 겨울은 기온이 낮고 대기가 건조하고 일조량이 적어지는 계절이다. 이런 겨울철에는 신선한 채소를 구하기 어려웠기 때문에 이를 대비하기 위한 특별한 지혜가 필요했을 것이다.

초기에는 단순히 소금이나 장에 채소를 절이는 방식의 원시적인 형태의 김치의 기록이 보인다. 염장은 세계 어느 지역에서나 널리 행해져 온 가장 보편적인 식품 보존법의 하나다. 또 채소 절임이라는 형태의 음식도 세계 각지에서 다양한 방식으로 만들어 먹었다. 우리나라와 인접한 중국이나 일본에서도 장과 식초를 이용하거나 쌀겨와 술지게미 등을 이용해서 만든 파오차이나 쓰케모노 같은 채소 절임 요리를 먹어왔다.

우리나라에서는 고유의 식문화 속에서 우리의 풍토와 입맛에 맞게 독자적인 방식과 기술로 채소 저장식이 발전해 왔다. 채소를 소금에 절여서 보존성을 높이는 데서 그치지 않고 곡류를 넣어 젖산균의 먹이가 되는 당분을 더함으로써 새로운 발효가 일어나도록 하였다. 젖산균이 채소 속의 당분을 분해하여 젖산을 비롯한 유기산을 만들어내기 때문에 김치는 특유의 새콤한 맛과 탄산감을 가지게 되었고 이는 우리 김치만의 고유한 특성이 되었다.

발효는 미생물이 유기물을 분해하여 화학적 변화를 일으키는 현상이라는 점에서 그 근본적인 원리는 부패와 다르지 않다. 다만 우리에게 이로운 부산물이 생성되어 맛과 향, 영양, 식감이 좋아지는 경우를 우리는 발효라고 부른다. 음식이 부패하지 않고 발효가 잘 이루어지도록 하려면 미생물의 활동을 적절하게 통제할 수 있어야 한다. 그래서 대부분의 채소 절임은 발효를 유도하는 종류의 음식이라고 하더라도 재료의 구성이 단순하고 소금이나 산 등 음식의 저장성을 높이는 재료의 비중이 높아진다.

그런데 한국의 김치는 발효를 시키기 위해 첨가하는 양념에도 다채로운 풍미를 내기 위한 다양한 재료를 넣는다는 점이 아주 독특하다. 이 양념에는 파와 마늘, 고추 같은 향신채도 들어가고, 생선이나 젓갈 같은 동물성 재료도 포함되어서 김치의 짠맛과 신맛, 단맛, 매운맛 그리고 깊은 감칠맛까지 만들어낸다.

한편 한국 음식은 전통적으로 주식과 부식의 뚜렷한 구분이 있어왔다. 우리나라는 일찍부터 벼가 주요한 작물이었고 쌀밥을 주식으로 하는 음식 문화가 오랫동안 이어져 왔다. 그러면서 주식인 쌀밥을 먹기 위해 밥에 곁들여서 밥맛을 돋우고 영양분을 보충할 수 있게 하는 음식인 반찬이라는 개념이 발달했다. 반찬은 단순히 주식에 곁들이는 음식이 아니라 밥 한 숟갈마다 곁들여져 맛을 완성하는 요소라는 점에서 서양의 '사이드 디시'와는 다른 개념이다.

김치는 우리 밥상에서 빠지지 않는 가장 중요한 반찬이다. 전통 반상에서도 김치는 국, 장과 함께 반드시 상에 올라야 하는 음식으로서 부식의 가짓수를 셀 때에도 포함시키지 않을 정도로 중요하게 취급되었다. 오늘날 여러 다른 음식 문화의 영향을 받아 식생활의 양식이 다양해지고 있는 가운데 여전히 김치는 상차림에서 기본이 되는 음식으로서 그 중요성을 잃지 않고 있다. 밥과 김치로 이루어진 이 기본적인 반상 차림의 구조가 오늘날까지도 우리 일상에서 유지되고 있다.

한국인에게 김치가 가지는 의미는 크다. 그만큼이나 우리가 먹는 김치의 종류도 아주 다양하다. 밥상에 여러 가지 김치가 한꺼번에 올라오는 일도 자연스럽다. 김치는 다양한 양념 재료를 넣어서 발효와 숙성이 이루어지게 한다는 점에 그 특별함이 있는 음식이다. 그래서 주재료가 어떤 채소가 되든 그 모두를 김치라고 부르며 김치의 종류는 세상에 알려진 채소의 가짓수만큼 많을 수 있다.

김치는 주로 겨울철에도 채소를 먹기 위해 만드는 저장 음식이었지만, 여름철에 쉽게 무르는 채소를 더 맛있게 먹기 위해 만드는 음식이기도 했다. 그 계절에 나는 채소로 담가 먹는 다양한 계절 김치도 오랜 역사를 가지고 있다. 봄과 여름에는 산나물과 열무, 오이 같은 채소로 만든 신선한 김치를 즐기고, 가을과 겨울에는 저장성이 뛰어난 배추와 무로 김장을 담갔다. 또 서해안에서는 새우젓, 남해안에서는 멸치젓, 내륙 지역에서는 상대적으로 담백한 양념이 사용되는 등 지역마다 선호하는 젓갈과 양념의 종류가 달랐다. 기온이 높은 남부 지역의 김치는 너무 빨리 익어버리지 않도록 소금과 양념, 젓갈을 많이 넣어서 진한 맛을 내고, 북부 지역의 김치는 양념은 적고 국물은 넉넉하게 만들어 시원한 맛을 냈다. 이처럼 각 지역의 환경과 생활 문화를 반영하면서 지역마다 서로 다른 특색을 지니는 다양한 김치가 만들어졌다.

　　과거에는 겨울을 대비하기 위해 많은 양의 김치를 한꺼번에 담가야 했고 이를 '김장'이라고 불렀다. 김장에 대한 기록은 고려 시대 사료에서부터 남아 있으며 조선 시대에 이르면서 전국적으로 널리 행해졌다고 알려진다. 김장은 가족과 친척, 마을 사람들이 함께 모여 겨우내 먹을 가장 중요한 반찬을 준비하는 행사로 협동과 나눔, 공동체의 유대를 상징하는 중요한 문화적 전통이었다. 해마다 같은 계절에 반복해서 행해지는 전통으로서 여러 세대에 걸쳐서 삶의 지혜를 전수하는 장이 되기도 했다. 이러한 가치를 인정받아 2013년에는 유네스코 인류무형문화유산으로 등재되기도 하였다. 이제는 더 이상 김장이 마을 공동체의 행사로 수행되고 있지는 않지만, 수행하는 단위가 축소되었을 뿐 김장은 계속해서 새로운 문화 관습을 만들어내며 오늘날까지 이어지고 있는 전통이다.

　　김치냉장고 같은 새로운 기술들이 보급되면서 더 이상 월동 준비를 위해 연례행사로 김장을 할 필요가 없어졌다. 통신 및 유통 수단이 발달하면서 김치의 지역색이 옅어지고 맛과 제조법의 표준화도 어느 정도 이루어졌다. 이제는 김치 그 자체를 반찬으로 먹는 데 머무르지 않고 김치를 재료로 쓰는 다양한 요리가 보편화되었다. 이러한 변화들을 맞이하면서도 김치와 김장에 관련된 관습들은 여전히 계속되고 있다. 김치는 강력한 고유성과 성체성을 지닌 음식이지만, 동시에 전통의 경계를 넘어서 다양한 식문화와 어우러질 수 있는 유연성과 확장성도 갖추고 있다.

현재 우리가 만들어 먹고 있는 김치의 재료나 담그는 법, 형태도 사회·경제·문화적 조건이나 기술의 변동, 사람들의 기호 등을 유연하게 받아들이면서 지금의 모습에 가까워져 온 것이다. 오늘날에는 고춧가루를 넣은 배추김치가 가장 일반적이지만, 김치에 고추를 넣어서 만들기 시작한 것은 조선 후기 무렵부터로 역사가 그리 길지 않다. 배추를 본격적으로 이용하기 시작한 것도 18세기부터라고 알려져 있으며, 지금과 같은 품종의 결구배추가 널리 재배된 것은 또 더 나중의 일이다. 배추김치도 다양한 변화를 겪으면서 지금과 같은 통배추김치의 형태로 자리 잡았다고 한다. 김치는 실로 살아 있는 문화라고 할 수 있다.

이 책은 계절별로 제철 채소를 이용해 담가 먹을 수 있는 김치의 종류와 그 조리법을 소개하고 있다. 우리 땅에서 계절마다 이렇게 다양한 채소가 나고 있고, 이 채소들마다 어울리는 다양한 김치 담그는 법이 전해져 있어서 모든 계절에 걸쳐서 다채롭게 김치를 즐길 수 있다는 점을 알려주고 싶었다. 가지처럼 오래전에는 김치의 주재료였지만 이제는 김치로 잘 담가 먹지 않는 재료를 이용해서 만드는 김치부터, 토마토나 연근처럼 김치의 재료로는 생소한 채소를 이용한 김치, 젓갈 없이 간장으로 담가 먹는 장김치, 해안 지역에서 풍부하게 생산되는 해산물에 무와 곡물을 곁들여 만들어 먹었던 식해 담그는 방식을 적용한 깍두기, 봄과 여름에 나는 채소로 무쳐서 바로 먹을 때 가장 맛이 좋은 김치들부터 오래 숙성시키면서 맛을 들여 먹는 김치들까지. 내 어머니로부터 전해진 김치들과 내가 좋아하는 다양한 계절의 맛들을 담아내기 위해 고민하고 연구하면서 발전시킨 김치 조리법까지이 책에 모두 소개하였다.

The Story of Kimchi Through
the Seasons

Kimchi is one of Korea's signature traditional foods rooted in the ancient custom of salting and processing vegetables for preservation. On the Korean Peninsula, where the seasons are distinct, people developed ever more ingenious ways to keep vegetables edible and flavorful year-round. In the hot, humid summer months, freshly harvested produce spoiled quickly, making preservation essential. Winters, meanwhile, were cold, dry, and dark, when it was difficult to obtain fresh vegetables at all. Preparing for these conditions required both skill and experience.

In the earliest records, kimchi appears in simple forms: vegetables salted or preserved in soy sauce. Salting is one of the world's most widespread methods of food preservation, and versions of pickled vegetables can be found in many cultures. China and Japan, too, have long traditions of pickled vegetable dishes such as *paocai* and *tsukemono*, made with soy sauce or vinegar, or fermented with rice bran or rice wine lees.

In Korea, these preservation practices developed into kimchi: a distinctive tradition shaped by local tastes, climate, and culinary knowledge. Rather than relying on salt alone, Koreans began adding grains. As these break down into sugars, they feed lactic acid bacteria and encourage fermentation. The bacteria transform the vegetables' natural sugars into lactic acid and other organic acids, giving kimchi its distinct tang and gentle effervescence that have become hallmarks of Korean kimchi.

Fermentation is a controlled transformation. Like spoilage, it involves microorganisms breaking down organic matter, but when conditions are properly managed, the process produces desirable flavors, aromas, textures, and nutritional benefits. For fermentation to proceed in the right direction, the activity of these microorganisms must be carefully controlled. This is why many pickled vegetables, even those intended for fermentation, typically rely on simple ingredient lists and relatively high levels of salt or acidity to ensure safe preservation.

Korean kimchi, however, stands out for the range and richness of ingredients used in the seasoning mixture that drives fermentation. Along with aromatic vegetables such as spring onions, garlic, and red peppers, many kimchi recipes include seafood elements like fish or salted and fermented seafood (*jeotgal*). Together, these ingredients create kimchi's complex balance of flavors: salty, sour, sweet, spicy, and deeply savory.

Korean meals have traditionally centered on a staple food (*jusik*) — most often rice — accompanied by side dishes (*busik*). This led to the development of the concept of *banchan*, side dishes that are eaten with rice to enhance its flavor and provide additional nutrients. *Banchan* is different from a Western "side dish" because it's not just an accompaniment; it's an essential element that completes the taste of every spoonful of rice.

Kimchi is the most essential *banchan* on a Korean table. In traditional meal settings, it was considered so fundamental that it was always served alongside soup and fermented sauces (*jang*) and was sometimes not even counted among the other side dishes. Even as Korean dining habits have diversified, the simple pairing of rice and kimchi remains a daily foundation.

Just as kimchi holds great meaning for Koreans, the variety of kimchi is remarkably broad. It is common for several kinds of kimchi to appear on the table at once. What makes kimchi special is not one single vegetable, but the method: vegetables are seasoned, then allowed to ferment and mature into something greater than the sum of their parts. In that sense, almost any vegetable can become kimchi, and the number of possible varieties is virtually endless.

Kimchi began as a way to store vegetables for the winter, but it also developed as a way to enjoy vegetables that soften quickly in summer heat. Seasonal kimchi made from what was available at different times of the year has a long history. In spring and summer, people enjoyed fresh kimchi made with wild greens, young summer radish, or cucumbers. In autumn and winter, they prepared *gimjang* (the tradition of making large quantities of kimchi for the colder months) using vegetables that store well, especially *baechu* (also known as Napa cabbage or Chinese cabbage) and radishes.

Regional tastes also shaped kimchi. On the west coast, salted shrimp sauce (*saeujeot*) was commonly used; along the south coast, salted anchovy sauce (*myeolchijeot*) was preferred; inland regions often favored milder seasoning. In Korea's warmer southern areas, kimchi tended to be seasoned more boldly, with extra salt and salted seafood sauce to prevent it from fermenting too quickly. In colder northern regions, kimchi was often made with lighter seasoning and more brine, resulting in a cleaner, more refreshing taste. Over time, local environments and ways of life gave rise to countless regional styles of kimchi.

Historically, preparing kimchi for winter meant making it in large batches all at once — an event known as *gimjang*. Records of *gimjang* date back to the Goryeo Dynasty, and it became widely practiced across the country during the Joseon Dynasty. Families, relatives, and neighbors gathered to prepare the most important side dish of the winter together, making *gimjang* a tradition closely associated with cooperation, sharing, and community bonds. Repeated each year in the same season, it also became a way for generations to pass down practical knowledge and family taste. In 2013, *gimjang* was inscribed on UNESCO's Representative List of the Intangible Cultural Heritage of Humanity. While it is no longer a village-wide communal event in many places, it has not disappeared. Instead, it continues, albeit on a smaller scale, adapting to modern life while remaining a living tradition.

As new technologies like kimchi refrigerators became widely available, making kimchi for winter no longer needed to be an annual necessity. With faster communication and distribution networks, regional differences in kimchi have become less pronounced, and certain flavors and recipes have become more standardized. Kimchi is now enjoyed not only as a side dish but also as an ingredient in a variety of popular dishes. Even so, the customs surrounding kimchi and *gimjang* continue.

Kimchi carries a strong sense of identity, yet it is also remarkably versatile — able to evolve, absorb new influences, and find its place in changing food cultures. The ingredients, methods, and forms of the kimchi we eat today have evolved in response to socioeconomic and cultural conditions, technological changes, and shifting tastes. Although *baechu*-kimchi with red pepper powder is now the most familiar variety, red peppers were added to kimchi only from the late Joseon period, so the history is not as long as many assume. *Baechu* itself became widely used relatively late, beginning around the 18th century. The modern variety of heading-type *baechu* was cultivated even later. Over time, *baechu*-kimchi changed and adapted, eventually taking shape as the whole-head *baechu*-kimchi we know today. In this sense, kimchi truly is a living culture.

In this book, I introduce a variety of kimchi types and recipes that can be made with seasonal vegetables. I wanted to show the richness of Korea's seasonal produce from its soil, as well as the many ways people have traditionally matched different vegetables with different kimchi-making methods so that kimchi can be enjoyed in every season of the year. These pages include recipes made with ingredients such as eggplant, once a common kimchi vegetable but less often used today, as well as newer combinations using tomatoes or lotus root. You will also find soy sauce-based kimchi made without salted seafood sauce, as well as *kkakdugi* (diced white radish kimchi) inspired by coastal *sikhae* traditions that ferment white radish and grains with abundant local seafood. From kimchi best eaten fresh, such as those made with spring and summer vegetables, to kimchi that deepens in flavor through long aging, this collection brings together recipes passed down from my mother as well as new recipes I developed through research — each one reflecting the seasonal flavors I love.

봄

SPRING

김치

KIMCHI

01

고수김치

Gosu-Kimchi

Coriander Kimchi

고수는 향이 좋아 향채(香菜)로도 불린다. 중국이나 동남아 지역에서 많이 사용하는 것과 달리 우리나라에서는 주로 사찰에서 사용하던 채소다. 해외여행이 자유로워지면서 이제는 누구나 즐겨 먹는 일반적인 식재료로 널리 애용되는 추세다.

고수김치는 봄에 담그는 것이 맛있다. 고들빼기나 냉이처럼 두해살이 식물이라 겨울을 이긴 고수를 뿌리까지 김치로 담그면 겉절이 상태로 먹었을 때 잎의 향과 뿌리의 고소함이 어우러져 그 진가를 발휘한다. 고수김치는 밥반찬으로 그냥 먹어도 좋고 육류를 먹을 때 쌈 채소 대신 먹어도 좋다.

Coriander is also called hyangchae (香菜) due to its pleasant aroma. Unlike its widespread use in China and Southeast Asia, coriander was mainly used in Korean temples. However, with increased travel opportunities, it is now widely enjoyed as a common ingredient.

Spring is the best time for making coriander kimchi. As a biennial plant like *godeulppaegi* (bitter lettuce) or *naengi* (shepherd's purse), coriander that has endured the winter is used with its roots to make kimchi. When eaten fresh, the aroma of the leaves and the nuttiness of the roots combine to show the real worth of the kimchi. Coriander kimchi is great as a side dish or can be enjoyed instead of lettuce wraps when eating meat.

재료

고수 500g, 부추 100g, 양파 1개(중간 이하 크기)

김치 양념: 쌀풀 1컵(쌀가루 1큰술, 물 1컵), 멸치 액젓 3큰술, 새우젓 3큰술, 고춧가루 1컵, 통깨 2큰술, 배즙 1컵(2쪽 일러두기 참고), 생강즙 1작은술

만드는 법

1 고수는 다듬어 깨끗이 씻은 후 먹기 좋은 크기로 썬다.

2 부추는 다듬어 씻은 후 먹기 좋은 크기로 썬다.

3 양파는 껍질을 벗기고 씻어 곱게 채 썬다.

4 쌀가루를 물에 잘 풀어 불에 올리고 저으면서 풀을 쑤어 식힌다.

5 김치 양념 재료를 모두 잘 섞는다.

6 양념에 먹기 좋게 썬 고수, 부추, 양파를 넣고 고루 버무린다.

7 겉절이로 맛있게 먹고, 남은 김치는 숙성시켜 먹는다.

Ingredients

Coriander 500g, chives 100g, onion 1 (medium or smaller)

Kimchi seasoning: nonglutinous rice paste 1 cup (nonglutinous rice powder 1 tablespoon, water 1 cup), salted anchovy sauce 3 tablespoons, salted shrimp sauce 3 tablespoons, red pepper powder 1 cup, roasted sesame seeds 2 tablespoons, pear juice 1 cup, ginger juice 1 teaspoon.

Instructions

1 Trim and wash the coriander, then cut it into bite-sized pieces.

2 Trim and wash the chives, then cut them into bite-sized pieces.

3 Peel and wash the onion, then shred finely.

4 Dissolve the nonglutinous rice powder in water, boil, and stir to make a paste. Let it cool.

5 Mix all the kimchi seasoning ingredients well.

6 Mix the cut coriander and chives along with the shredded onion and seasoning.

7 Enjoy some of it as *geotjeori* (fresh kimchi) and let the rest ferment to enjoy later.

고수는 우리나라에서는 주로 사찰에서 사용하던 채소였으나
이제는 누구나 즐겨 먹는 일반적인 식재료로 널리 애용되는 추세다.
Coriander was primarily used in Korean temples, today, it is widely
enjoyed as a staple ingredient.

02

참죽순김치

Chamjuksun Kimchi
Toona Sinensis Shoots Kimchi

참죽나무의 연한 어린잎을 참죽순이라 부른다. 독특한 향을 가지고 있으며 고기 맛이 나는 매력적인 식재료다. 사찰에서는 억센 줄기와 잎을 말려두었다가 국물을 내는 데 사용하기도 한다. 한번 맛을 본 사람은 해마다 봄이 되면 저절로 찾게 되는 중독성이 강한 식물이다. 막 담가서 바로 먹는 겉절이는 아삭하니 씹는 식감과 함께 향미가 도드라진다. 익히면 개성 있는 향미가 깊어져 좋고 김장 김치처럼 묵혀서 먹으면 또 다른 깊은 맛을 느낄 수 있다.

신선하지 않은 참죽순은 손으로 잡고 흔들면 잎이 모두 떨어지고 줄기만 남는다. 고추장과 잘 어울려 고추장장아찌, 고추장부각, 장떡 같은 음식으로 해 먹기도 한다.

* 가죽나무와 참죽나무는 다르다. 잎을 먹는 건 참죽나무로 봄에 어린순을 따서 나물, 김치, 전, 부각 등을 해서 먹는다.

* *Gajuk* tree (Ailanthus altissima Swingle) and *chamjuk* tree are not the same. The edible leaves come from chamjuk tree, and its young spring leaves are known as chamjuksun. In spring, people harvest the tender shoots and use them in seasoned greens (*namul*), kimchi, pan-fried delicacies (*jeon*), and deep fried-vegetable coated with starch(*bugak*).

Chamjuksun are the tender young shoots of the *chamjuk* tree (Toona sinesis or Chinese Mahogany, Chinese Toon). They have a unique aroma and a meat-like flavor, making them an attractive ingredient. In temples, the tough stems and leaves are dried and used for broth. Once tasted, *chamjuksun* becomes addictive, drawing enthusiasts every spring. Freshly made *geotjeori* (fresh kimchi) is crispy with a satisfying texture and flavor. When fermented, its unique aroma is enhanced and deepened, and as it ages, its aroma becomes deeper with a variety of flavors, like aged kimchi.

Chamjuksun pairs well with red pepper paste (*gochujang*) and can be used in a variety of dishes such as *chamjuksun gochujang jangajji* (pickled

chamjuksun in *gochujang*), *chamjuksun gochujang bugak* (deep-fried *chamjuksun* coated with starch and *gochujang*), and *chamjuksun jjangddeok* (pan-fried *chamjuksun gochujang* patty).

재료

손질한 참죽순 400g, 5% 소금물 1L(물 950ml, 소금 50g)

김치 양념: 멥쌀 풀 1컵(참죽순 끓인 물 1컵, 멥쌀가루 1큰술), 고춧가루 3~4큰술, 멸치 액젓 2큰술, 쪽파 5뿌리, 다진 마늘 1큰술

만드는 법

1   참죽순을 다듬어 씻어 5% 소금물에 1~2시간 절인다.
2   참죽순을 다듬고 남은 거친 잎과 줄기들을 물에 넣고 끓인 뒤 식힌다.
3   절인 참죽순을 물에 2~3번 헹궈 건져 물기를 뺀다.
4   2의 참죽순 끓인 물에 멥쌀가루를 넣고 풀을 쑤어 식힌다.
5   풀이 식는 동안 쪽파를 다듬어 씻어 송송 썬다.
6   멥쌀 풀에 고춧가루와 멸치 액젓, 송송 썬 쪽파, 다진 마늘을 넣고 잘 섞어 김치 양념을 만든다.
7   물기를 뺀 참죽순에 6의 김치 양념을 넣고 잘 버무린다.
8   양념이 묻은 그릇에 물을 1컵 넣고 부셔서 김치 국물로 쓴다.
9   겉절이로 먹다가 익혀 먹어도 좋다.

Ingredients

Trimmed *chamjuksun* (*Toona sinensis* shoots) 400g, 5% salt water 1L (water 950ml, salt 50g)

Kimchi seasoning: nonglutinous rice paste 1 cup(*chamjuksun*-blanched water 1 cup, nonglutinous rice powder 1 tablespoon), red pepper powder 3-4 tablespoons, salted anchovy sauce 2 tablespoons, green

Instructions

1   Trim the tough shoots of the *chamjuksun*. Wash the trimmed *chamjuksun* and soak in 5% salt water for 1-2 hours.
2   Boil the tough shoots and stems of the *chamjuksun* left over from trimming in water.
3   Rinse the salted *chamjuksun* 2-3 times in water, then drain.
4   Meanwhile, put the nonglutinous rice powder in the *chamjuksun* blanched water from the Step 2 and boil to make the nonglutinous rice paste.
5   While the paste cools, trim, wash, and chop the green onions.

onions 5, minced garlic
1 tablespoon

6   Mix the nonglutinous rice paste with red pepper powder, salted anchovy sauce, green onions, and minced garlic to make the kimchi seasoning.

7   Blend the drained *chamjuksun* with the kimchi seasoning from Step 6 and mix well.

8   Add 1 cup of water to the bowl used for mixing the kimchi, stir to dissolve any remaining seasoning, and use this as kimchi liquid.

9   This kimchi can be enjoyed fresh as *geotjeori* (fresh kimchi) or after fermentation.

03

미나리김치

Minari Kimchi
Water Parsley Kimchi

묵은 김장 김치를 겨우내 먹다 보면 물려서 젓가락이 안 가는 시기가 온다. 이때 미나리로 김치를 담그면 봄 미나리의 신선한 향이 잠자던 식욕을 불러온다. 먹기 좋게 3~4cm 길이로 잘라서 담기도 하지만 부추와 길이를 맞춰 길게 담가 먹어도 좋다.

　겨우내 움츠리고 있던 몸에 쌓인 노폐물을 빼주는 데 좋은 미나리에 연근을 갈아 죽을 쑤거나 생으로 넣고 김치를 담그면 연근이 내는 구수함이 더해져 풍성한 맛의 미나리김치가 된다.

After eating aged *gimjang* kimchi all winter, there comes a time when you don't want to eat kimchi any longer. Making water parsley kimchi at this time will awaken your dormant appetite with the fresh scent of spring water parsley. Typically cut into 3-4cm lengths for easy eating, water parsley kimchi is also enjoyable when made with longer pieces, matching the length of the chives.

Water parsley is excellent for cleansing the body of toxins accumulated during winter. You can make porridge with ground lotus roots and water parsley or make parsley kimchi with raw lotus roots. The savory flavor of the lotus root will enhance the richness of the water parsley kimchi.

재료

미나리 300g, 부추 100g

김치 양념: 멸치 액젓 3큰술,
양파즙 1컵, 고춧가루 5큰술,
통깨 1큰술, 쪽파 5뿌리,
다진 마늘 1작은술, 생강즙
1작은술

만드는 법

1 미나리는 뿌리와 잎을 떼고 다듬어서 깨끗이 씻은 후 10분간 따뜻한
  물에 담가 두었다가 건져 물기를 뺀다.

2 부추는 깨끗이 씻는다.

3 미나리와 부추를 4cm 길이로 썬다.

4 양파는 갈아서 즙을 낸다.

5 쪽파는 뿌리와 누런 부분을 제거하고 깨끗이 씻어 4cm 길이로 썬다.

6 마늘은 다진다.

7 김치 양념 재료를 모두 넣고 고루 섞는다.

8 썰어 놓은 미나리와 부추를 김치 양념에 넣고 고루 버무린다.

9 겉절이로 먹다가 익혀 먹어도 좋다.

Ingredients

Water parsley 300g,
chives 100g

Kimchi seasoning:
salted anchovy sauce
3 tablespoons, onion
juice 1 cup, red pepper
powder 5 tablespoons,
roasted sesame seeds
1 tablespoon, green
onions 5, minced garlic 1
teaspoon, ginger juice 1
teaspoon

Instructions

1 Remove the leaves and roots from the water parsley, then
  trim and wash it. Soak in warm water for 10 minutes and
  drain.

2 Wash the chives.

3 Cut the water parsley and chives into 4cm lengths.

4 Grind the onion and extract the juice.

5 Remove the roots and yellow parts from the green onions,
  wash, and drain, then cut them into 4cm lengths.

6 Mince the garlic.

7 Mix all the seasoning ingredients well.

8 Add the cut water parsley and chives to the kimchi
  seasoning and mix all evenly.

9 Enjoy it fresh or let it ferment.

미나리로 김치를 담그면 봄 미나리의 신선한 향이 잠자던 식욕을 불러온다.
Making water parsley kimchi will awaken your dormant appetite with
fresh scent of spring water parsley.

04

미나리물김치

Minari Mul-Kimchi
Water Parsley Kimchi in Water

향이 좋은 미나리는 김치를 담그는 데 빠지면 서운한 조연급 배우다. 그런 미나리를 주연 삼아 김치를 담근다. 국물 없이 담그는 미나리김치도 좋지만 물김치로 담가 먹으면 시원한 국물이 미나리의 향을 담고 입으로 들어온다. 향만 좋은 것이 아니라 미나리의 속이 비어서 그렇겠지만 아삭하게 씹히는 특별한 식감을 즐기기에 부족함이 없다.

돌나물이나 나박나박 썬 무 등을 함께 넣고 담그기도 한다.

Water parsley, with its fragrant aroma, is a supporting actor that would be missed if not included in kimchi. This recipe features water parsley as the main ingredient. If you typically enjoy kimchi without liquid, try making it into water kimchi; the refreshing kimchi liquid carries the aroma of water parsley to your mouth. Not only does it have a delightful aroma, but it also has a unique crunchy texture, likely due to its hollow stems, making it enjoyable to eat.

You can also add ingredients like *dolnamul* (sedum) or thinly sliced white radish to this kimchi.

재료

미나리 500g, 쪽파 10뿌리, 마늘 5쪽, 생강 1쪽, 실고추 약간

김치 국물: 고춧가루 2큰술, 밀가루 풀 2컵(밀가루 2큰술, 멸치 육수 2컵), 배즙 1컵, 1.5% 소금물 3L(물 2,955ml, 소금 45g)

만드는 법

1 미나리는 잎을 모두 따내고 깨끗이 씻어 4~5cm 길이로 썬다.
2 쪽파는 다듬어 씻어 미나리와 같은 길이로 자른다.
3 마늘과 생강은 껍질을 벗기고 씻어 채 썬다.
4 멸치 육수에 밀가루를 잘 풀어 풀을 쑤어 식힌다.
5 준비한 김치 재료들을 고루 섞어 김치통에 담는다.
6 1.5% 소금물에 배즙과 밀가루 풀을 넣고 잘 섞는다.
7 고춧가루를 면포에 잘 싸서 6의 김치 국물에 넣고 흔들어 고추 색과 맛이 우러나게 한다.
8 5의 김치 재료가 든 용기에 7의 김치 국물을 부어 상온에서 익힌다.
9 국물이 80% 정도 익으면 낮은 온도의 냉장고에 넣어두고 천천히 익혀 먹는다.

Ingredients

Water parsley 500g, green onions 10, garlic 5 cloves, ginger 1 piece, red pepper threads (a small amount)

Kimchi liquid: red pepper powder 2 tablespoons, flour paste 2 cups (flour 2 tablespoons, anchovy broth 2 cups), pear juice 1 cup, 1.5% salt water 3L (water 2,955ml, salt 45g)

Instructions

1 Remove all the leaves from the water parsley, wash, and cut it into 4-5cm lengths.
2 Trim and wash the green onions, and cut them into the same length as the water parsley.
3 Peel, wash, and shred the garlic and ginger.
4 Dissolve the flour in the anchovy broth, boil to make a paste, and let it cool.
5 Mix the prepared kimchi ingredients well and place them in a kimchi container.
6 In 3L of 1.5% salt water, add the pear juice and flour paste, and mix well.
7 Wrap the red pepper powder tightly in a cotton cloth, then shake it in the kimchi liquid from Step 6 to release the color and flavor of the red pepper powder.
8 Pour the kimchi liquid from Step 7 over the ingredients in the kimchi container from Step 5, and let it ferment at room temperature.

9   When the kimchi liquid is 80% ripe, store it in the
    refrigerator at a low temperature, and let it ripen slowly
    before eating.

05

달래김치

Dallae Kimchi
Wild Chive Kimchi

이른 봄에 산에 가면 나무 그늘이나 물가 옆에 일찍 올라온 달래가 무더기로 자라고 있는 걸 본다. 호미로 넓게 캐서 줄기를 잡고 흔들면 흙은 떨어지고 하얀 뿌리들이 모습을 드러낸다. 다듬어 썰어 전을 부치기도 하고 달래간장을 만들어 밥도 비빈다.

산에서만 달래가 자라는 것은 아니다. 해마다 달래를 캐는 들판이나 밭으로 가서 호미로 땅을 헤집는다. 아직 싹은 올라오지 않았는데 땅속에 달래의 하얀 뿌리가 마늘처럼 들어 있다. 뽀얀 뿌리만 있는 달래라 해서 '은달래'라 불린다. 요즘 은달래는 모래 속에 심어 체로 쳐서 팔려 나오기도 한다.

김치로 담그면 더 맛있는데, 이때 달래는 줄기가 굵고 알도 큰 것이 좋다. 묵혀서 먹으려면 말할 것도 없이 알이 굵고 나이를 먹은 달래를 써야 한다. 갓 담근 달래김치는 알싸하니 향이 좋다가 푹 익은 후에는 쉽게 잊히지 않게 달고 깊은 맛으로 우리를 유혹한다.

In early spring, if you go to the mountains, you can see clusters of early wild chives growing under the shade of trees or near water. Using a hoe, you can dig them up, shake off the soil, and reveal the white roots. These can be cleaned and used to cook *jeon* (pan-fried delicacies) or wild chives soy sauce to mix with rice.

Wild chives don't only grow in the mountains. Every year, people go to fields and farms to dig them up using a hoe. Wild chives have yet to sprout, but their white wild chive roots are hidden underground like garlic cloves. These chives are called "silver wild chives" due to their white roots and are often grown in sandy soil and sold after sifting through a mesh.

The best wild chives for making kimchi are those with thick stalks and large bulbs. For aged kimchi, it's essential to use thick, mature wild chives. Freshly made wild chive kimchi has a piquant and fragrant taste, but as it ferments, it develops a sweet and deep flavor that is hard to forget.

재료

달래 500g, 쪽파 300g

김치 양념: 밀가루 풀 1컵(물 1컵, 밀가루 1큰술), 멸치 액젓 6큰술, 고춧가루 1컵, 통깨 2큰술, 양파즙 1컵

만드는 법

1   달래는 다듬어서 깨끗하게 씻어 물기를 뺀다. (잘게 썰어 담가도 좋다.)
2   쪽파는 다듬어서 깨끗하게 씻어 물기를 뺀다. (잘게 썰어 담가도 좋다.)
3   밀가루를 물에 잘 풀어 불에 올리고 저으면서 풀을 쑤어 식힌다.
4   김치 양념 재료들을 한데 넣어 고루 섞는다.
5   물기를 뺀 달래와 쪽파에 4의 양념을 넣고 버무린다.
6   처음엔 달래간장처럼 먹다가 익혀서 먹는다.

Ingredients

Wild chives 500g, green onions 300g

Kimchi seasoning: flour paste 1 cup (water 1 cup, flour 1 tablespoon), salted anchovy sauce 6 tablespoons, red pepper powder 1 cup, roasted sesame seeds 2 tablespoons, onion juice 1 cup

Instructions

1   Trim and wash the wild chives, then drain. (You can also chop the wild chives finely.)
2   Trim, wash the green onions, then drain. (You can also chop the green onions finely.)
3   Mix the flour with water and boil it while stirring to make the paste, then let it cool.
4   Combine all the kimchi seasoning ingredients together and mix well.
5   Add the kimchi seasoning from Step 4 to the drained wild chives and green onions, and mix thoroughly.
6   Initially, enjoy it like wild chive soy sauce, and then enjoy it as fermented kimchi.

06

좁쌀(기장)깍두기

Jopssal (Gijang) Kkakdugi

Millet - Diced White Radish Kimchi

보는 재미와 먹는 재미가 있는 김치가 좁쌀깍두기다. 이 김치의 시작은 가자미식해에서다. 가자미식해의 생선과 좁쌀, 무로 이어지는 맛을 젓갈, 좁쌀(기장), 무로 연결시켰을 뿐 다른 것은 없다.

　좁쌀이 삭으면서 유산균의 먹이로 쓰이고 남은 섬유질은 그 형태를 유지하는 힘이 된다. 그러니 익어도 좁쌀 알갱이가 살아 있어 식감이 독특하고 맛있는 김치라는 칭찬을 듣는다. 무를 절이지 않고 담기에 국물이 많은 김치인데 남은 국물에 밥 한 술 넣고 비벼 먹으면 다양한 김치를 먹을 수 있는 나라에서 사는 고마움이 절로 느껴진다.

Millet - diced white radish kimchi is visually appealing and delicious kimchi. It originates from *gajami sikhae* (spicy fermented sole). The taste of *gajami sikhae*, which combines fish, millet, and white radish, is linked to this kimchi, using salted fish sauce, millet, and white radish.

As millet rice ferments, it serves as food for lactic acid bacteria, and the remaining fiber helps to maintain its shape. Therefore, even when fermented, the millet grains remain intact, resulting in a unique texture and delicious kimchi that receives praise. This kimchi has plenty of kimchi liquid because the white radish is not salted beforehand. When I mix a spoonful of rice with the remaining liquid, I feel thankful for living in a country with such a variety of kimchi.

재료

무 2kg, 좁쌀(기장) 1컵,

김치 양념: 쪽파 100g, 멸치
액젓 1/2컵, 새우젓 1/2컵,
고춧가루 1컵, 다진 마늘
4큰술, 다진 생강 1작은술,
배즙 1컵

만드는 법

1   좁쌀(기장)은 깨끗하게 씻어서 잘 일은 다음 밥을 해서 식힌다.

2   무는 깨끗하게 씻어 사방 1.5cm 크기의 깍두기 모양으로 썬다.

3   썬 무에 멸치 액젓과 새우젓을 넣고 버무려 놓는다.

4   쪽파는 다듬어 씻은 후 1cm 길이로 썰고 마늘과 생강은 다진다.
    젓갈에 버무려 놓은 무에 고춧가루를 넣어 버무린다.

5   4에 식힌 좁쌀밥과 쪽파, 마늘, 생강, 배즙을 넣고 다시 버무린다.

6   상온에서 2~3일 정도 익힌 후 냉장고에 넣어두고 천천히 숙성시키
    면서 먹는다.

Ingredients

White radish 2kg, millet
1 cup

Kimchi seasoning: green
onions 100g, salted
anchovy sauce 1/2 cup,
salted shrimp sauce 1/2
cup, red pepper powder
1 cup, minced garlic 4
tablespoons, minced
ginger 1 teaspoon, pear
juice 1 cup

Instructions

1   Wash the millet, clean it using a straining ladle, and cook
    it into rice. Let it cool.

2   Wash the white radish and cut it into 1.5cm cubes.

3   Mix the diced radish with salted anchovy sauce and
    salted shrimp sauce and let it sit for salting.

4   Trim and wash the green onions, then cut them into 1cm
    lengths. Mince the garlic and ginger.
    Add the red pepper powder to the salted white radish and
    mix well.

5   Add the cooled millet rice, green onions, garlic, ginger,
    and pear juice to the mixture from Step 4 and mix again.

6   Let it ferment at room temperature for about 2-3 days,
    then store it in the refrigerator. Enjoy as it matures.

07

알배추물김치

Albaechu Mul-Kimchi

Baby *Baechu* Water Kimchi

알배추는 한 손에 잡히는 만만한 크기라 김치를 담그는 부담을 덜어주어 좋다. 붉은 고춧물을 내어 담가두면 익으면서 변해 가는 국물의 빛깔로 익은 정도를 파악할 수 있는데 '배추동치미'라 부르고 싶은 맛이다. 무는 국물이 시원하라고 넣지만 **빼도** 상관없다. 오래 저장해 두고 먹으려고 넉넉히 담근다면 국물용 소금물의 염도를 3%로 잡았다가 먹을 때마다 찬물을 부어 희석해 먹으면 좋다.

　배추를 자르지 않고 통으로 담그는 김치는 오래 천천히 익혀야 제맛이 난다. 빨리 익혀 먹고 싶다면 잘라서 담그기를 권한다.

As baby *baechu* is easy to handle with one hand, this makes the process of making kimchi less burdensome. If you soak it in red pepper water, you can determine the degree of ripeness by the changing color of the liquid as it ferments. It tastes like a briny water kimchi, made with white radish, called *dongchimi*, so I would call it *baechu dongchimi*. The white radish is added to make kimchi liquid refreshing, but it can be omitted. If you plan to store baby *baechu* water kimchi for a long time, use a 3% salt water solution and dilute it with cold water before eating.

Kimchi that is made with the whole *baechu* without being cut tastes best when fermented slowly for a long time. If you want to enjoy it sooner, I recommend cutting the *baechu* into pieces prior to seasoning.

* The type of cabbage used to make kimchi is also known as "Napa cabbage" or "Chinese cabbage". In this book, all varieties of kimchi cabbage (except for Western-style cabbage) are referred to as "*baechu*".

재료

알배추 3kg(물 2.7L, 소금 300g), 무 1kg(소금 20g), 대파 2뿌리, 마늘 8쪽, 생강 1쪽(마늘 크기 정도로 작은 것)

김치 국물: 2% 소금물 5L(물 4.9L, 소금 100g), 찹쌀 풀 3컵(물 3컵, 찹쌀가루 3큰술), 고춧가루 3큰술

만드는 법

1   알배추는 통째로 10% 소금물에 담가 하루 저녁 이상 절인다.
2   무는 깨끗이 씻어 먹기 좋은 크기로 썬 다음 소금을 뿌려 1시간 정도 절인다.
3   절인 알배추를 물에 2~3번 헹궈 물기를 뺀다.
4   찹쌀 풀을 쑤어 식힌다.
5   절인 무를 물에 헹궈 물기를 뺀다.
6   대파는 뿌리와 누렇게 변한 잎을 떼어내고 씻어 7cm 길이로 썬다.
7   마늘은 껍질을 벗겨 씻은 후 1mm 두께로 편 썬다.
8   생강은 껍질을 벗기고 씻은 후 1mm 두께로 편 썬다.
9   김치통에 무를 밑에 깔고 그 위에 3의 알배추와 대파, 마늘, 생강을 차곡차곡 담는다.
10   2% 소금물을 만든다.
11   10의 소금물에 식힌 찹쌀 풀을 섞는다.
12   고춧가루를 면포에 싸서 11의 소금물에 넣고 조물거리면서 고춧물을 뺀다.
13   김치통의 알배추 위로 12의 고춧물을 붓는다.
14   알배추나 무가 물 위로 떠오르지 않게 누르고 뚜껑을 닫는다.
15   상온에서 2~3일 숙성시킨 후 냉장고에 넣고 천천히 익혀 먹는다.

Ingredients

Baby *baechu* 3kg (water 2.7L, salt 300g), white radish 1kg (salt 20g), large spring onions 2, garlic 8 cloves, ginger 1 small piece (similar to the garlic in size)

Kimchi liquid: 2% salt water 5L (water 4.9L,

Instructions

1   Soak the whole baby *baechu* in 10% salt water overnight.
2   Wash the white radish, cut it into bite-sized pieces, sprinkle salt on it, and let sit for about an hour.
3   Rinse the salted baby *baechu* 2-3 times in water and drain.
4   Prepare the glutinous rice paste and let it cool.
5   Rinse the salted white radish and drain.
6   Remove the roots and wilted leaves of large spring onions

salt 100g), glutinous rice paste 3 cups(water 3 cups, glutinous rice powder 3 tablespoons), red pepper powder 3 cups

and wash, cut them into 7cm lengths.

7   Peel the garlic, wash and slice it into 1mm thick pieces.

8   Peel the ginger, wash and slice it into 1mm thick pieces.

9   In a kimchi container, place the white radish at the bottom, then place baby *baechu* from Step 3 on top of it.

10 Prepare 2% salt water.

11 Mix the cooled glutinous rice paste with the salt water from Step 10.

12 Wrap the red pepper powder in a cotton cloth and squeeze it into the salt water from Step 11 to make the red pepper water.

13 Pour the red pepper water from Step 12 over the baby *baechu* in the kimchi container.

14 Press down on the baby *baechu* and the white radish to prevent them from floating above the water and close the lid.

15 Let it ferment at room temperature for 2-3 days, then store to further ferment it slowly in the refrigerator. Enjoy once it has fermented.

산갓물김치

08

Sangat Mul-Kimchi
Wild Mustard Leaf Water
Kimchi

산갓은 '늦쟁이냉이'의 별칭이다. 재배하고 있는 곳이 없어 오로지 채집에만 의존해 구할 수 있는 식물인데, 고추냉이와 갓의 알싸한 맛을 가지고 있다. 아마 그래서 별칭이 산갓인 것으로 추정된다.

산갓물김치는 『음식디미방』(조선 후기에 장계향이 쓴 국내 최초의 한글 조리서)에 언급됐을 만큼 역사가 오래된 김치이지만 재료의 수급이 쉽지 않아 대중에게 널리 알려지지 않았다.

이른 봄 깊은 산속 습지에서 볼 수 있는 산갓을 채취해 물김치로 담그는데, 죽이나 떡 등과 함께 먹으면 소화에 도움을 주는 것은 물론이고 알싸한 매운맛으로 춘곤증을 잊게 하는 김치다. 재배가 가능한 때를 기다리는 김치다.

Sangat (wild mustard leaf) is another name for *Neunjaengi naengi*. This plant is not cultivated but rather can only be obtained by foraging and it has a pungent taste of wasabi and *gat* (mustard leaf). Maybe that's why this plant is also called wild mustard leaf.

Although this kimchi is old enough to be mentioned in the *Eumsikdimibang* (the first cookbook written in *Hangeul* by Jang Kyehyang during the late Joseon period), it is not widely known to the public due to the difficulty of obtaining the ingredients.

In early spring, you can gather the wild mustard leaf from the wetlands of deep mountains and make it into water kimchi. Eating this water kimchi with porridge or rice cake helps not only digestion but also alleviates spring fever with its bitter and spicy taste. This kimchi awaits the day it can be cultivated.

재료

산갓 500g, 무 200g, 쪽파 5줄기, 마늘 5쪽

김치 국물: 1.5% 소금물 5L(물 4,925ml, 소금 75g), 멥쌀 풀 1컵(물 1컵, 쌀가루 1큰술)

만드는 법

1 멥쌀 풀을 쑤어 식힌다.
2 1.5% 소금물을 만든다.
3 산갓은 잎이 상하지 않게 살살 씻어 건져 놓는다.
4 무는 2cm×1.5cm 크기로 얇게 썰고, 마늘도 얇게 편으로 썬다.
5 쪽파는 뿌리를 떼어내고 손질해 씻은 후 2cm 길이로 썬다.
6 소금물에 멥쌀 풀을 넣고 고루 잘 섞는다.
7 김치통에 산갓, 무, 마늘, 쪽파를 담고 6의 물을 부어 상온에 2~3일 두어 익힌 후 냉장 저장해서 2주 정도 숙성시켜서 먹는다.

Ingredients

Wild mustard leaf 500g, white radish 200g, green onions 5

Kimchi liquid: 1.5% salt water 5L (water 4,925ml, salt 75g), nonglutinous rice paste 1 cup (water 1 cup, nonglutinous rice powder 1 tablespoon)

Instructions

1 Make the nonglutinous rice paste and let it cool.
2 Prepare 1.5% salt water.
3 Wash the wild mustard leaves carefully to avoid damaging them, and drain.
4 Thinly slice the white radish into 2cm x 1.5cm pieces. Slice
5 the garlic thinly.
   Remove the roots of the green onions, trim and wash them, and cut into 2cm lengths.
6 Add the nonglutinous rice paste to the salt water and mix well.
7 Place the wild mustard leaves, white radish, and green onions in a container and add the salt water mixture from Step 6. Ferment it at room temperature for 2-3 days, then store it in the refrigerator for about 2 weeks to continue fermenting before eating.

09

봄동사과김치

Bomdong Sagwa Kimchi

Spring *Baechu*-Apple Kimchi

묵은 김치가 싫증 날 무렵에 담가 먹기 좋은 김치가 봄동사과김치다. 겨우내 땅의 기운을 받고 추위를 이기며 자란 봄동 한 포기와 막 올라온 달래에 먹다 남은 사과 1개를 준비한다. 봄동을 한입에 먹기 좋게 자른 뒤 사과도 봄동의 크기에 맞춰 썬다. 큰 기술 없이도 김치 양념만 넣고 버무려 겉절이로 먹어보면 왜 그동안 샐러드에 탐닉하며 살았나 하는 후회가 될 것이다.

Spring *baechu*-apple kimchi is a great alternative when you're tired of aged kimchi. Prepare a head of spring *baechu* that has grown through the winter, absorbing the earth's energy and enduring the cold, along with some freshly sprouted wild chives and a leftover apple. Cut the spring *baechu* into bite-sized pieces and cut the apple to match the size of the spring *baechu*. Making fresh kimchi by simply tossing it with the kimchi seasoning requires no special skills, and I wonder why I have been indulging in salads for so long.

재료

봄동 300g, 달래 50g, 사과 1개

김치 양념: 멸치 액젓
3~4큰술, 고춧가루 3큰술,
다진 대파 1큰술, 다진 마늘
1작은술, 식초 1작은술, 볶은
통깨 1큰술

Ingredients

Spring *baechu* 300g, wild
chives 50g, apple 1

Kimchi Seasoning:
salted anchovy sauce 3-4
tablespoons, red pepper
powder 3 tablespoons,
chopped large spring
onion 1 tablespoon,
minced garlic 1 teaspoon,
vinegar 1 teaspoon,
roasted sesame seeds 1
tablespoon

만드는 법

1   봄동은 깨끗이 씻어 손으로 먹기 좋게 자른다.

2   달래는 뿌리 쪽을 깨끗하게 다듬어 씻는다.

3   사과는 깨끗하게 씻어 껍질째 봄동과 비슷한 크기로 썬다.

4   볼에 멸치 액젓을 넣고 고춧가루와 다진 대파, 다진 마늘을 넣고 잘
    섞는다.

5   1의 봄동을 4의 무침 양념에 잘 버무린다.

6   버무린 봄동에 사과와 달래를 넣고 고루 섞는다.

7   식초와 볶은 통깨를 넣고 마무리한다.

Instructions

1   Wash the spring *baechu* and tear it into bite-sized pieces
    by hand.

2   Trim and wash the roots of the wild chives.

3   Wash the apple and cut it into the same size as the spring
    baechu, leaving the skin on.

4   In a bowl, combine the salted anchovy sauce, red pepper
    powder, chopped large spring onion, and minced garlic
    together and mix well.

5   Add the prepared spring *baechu* from Step 1 to the kimchi
    seasoning from Step 4 and mix thoroughly.

6   Add the apple and wild chives to the mixed spring *baechu*
    and mix well.

7   Finish by adding vinegar and roasted sesame seeds.

겨우내 땅의 기운을 받고 추위를 이기며 자란 봄동 한 포기와 막 올라온 달래로,
묵은 김치가 싫증 날 무렵 봄동사과김치를 담가 먹는다.

When you're tired of aged kimchi, spring *baechu*-apple kimchi is a great alternative.
It is made with a head of spring *baechu* grown through the winter — absorbing the earth's
energy and enduring the cold — along with some freshly sprouted wild chives.

10

상추겉절이

Sangchu Geotjeori
Lettuce Fresh Kimchi

상추 하면 흔히 쌈 채소라고 생각하지만, 상추로 전을 부치기도 하고 국도 끓이며 떡을 찌기도 한다. 쌈으로 먹다 남은 상추가 있다면 냉장고 속 자투리 채소들을 챙겨 넣고 겉절이로 무쳐 먹는다. 직전에 먹은 쌈과 다른 맛이라 놀랄 것이다.

상추, 이제는 쌈 말고 김치로도 담가 먹자고 추천한다.

When it comes to lettuce, people usually think of wraps, but it can also be used to make _jeon_ (pan-fried delicacies or pancakes), soup, and rice cakes. If you have leftover lettuce from enjoying wraps, you can add the remaining vegetables in the refrigerator and make them into _geotjeori_ (fresh kimchi). You will be surprised by the different taste compared to the wraps you just ate.

I recommend that you try making kimchi with lettuce instead of using it for wraps.

재료

손질한 상추 200g, 손질한
쑥갓 100g

김치 양념: 양파 1/4개, 실파
5~6뿌리, 새우 가루 1작은술,
통깨 1큰술, 참기름 2작은술,
실고추 약간, 고춧가루
2큰술, 다진 마늘 1작은술,
생강술 1큰술(2쪽 일러두기
참고), 사과즙 2큰술, 멸치
액젓 2큰술

만드는 법

1   상추는 연하고 싱싱한 것을 골라 다듬어 씻어 건진다.

2   쑥갓은 연하고 싱싱한 것을 골라 다듬어 씻어 건진다.

3   실파는 다듬어서 2~3cm 길이로 썬다.

4   양파는 곱게 채 썬다.

5   실고추는 실파와 비슷한 길이로 자른다.

6   마늘은 다진다.

7   김치 양념 재료를 고루 섞는다.

8   준비한 김치 양념에 채 썬 양파, 새우 가루, 통깨를 넣고 잘 섞는다.

9   물기를 뺀 상추와 쑥갓에 김치 양념을 넣고 살살 버무린다.

Ingredients

Trimmed lettuce 200g,
trimmed crown daisy
100g

Kimchi seasoning:
onion 1/4, thin spring
onions 5-6, shrimp
powder 1 teaspoon,
roasted sesame seeds 1
tablespoon, sesame oil
2 teaspoons, red pepper
threads (a small amount),
red pepper powder 2
tablespoons, minced
garlic 1 teaspoon, ginger
liquor 1 tablespoon,
apple juice 2 tablespoons,
salted anchovy sauce 2
tablespoons

Instructions

1   Select soft and fresh parts of the lettuce, trim, wash, and drain.

2   Select soft and fresh parts of the crown daisy, trim, wash, and drain.

3   Trim the thin spring onions and cut them into 2-3cm lengths.

4   Slice the onion finely.

5   Cut the red pepper threads to a similar length as the thin spring onions.

6   Mince the garlic.

7   Mix all the kimchi seasoning ingredients well.

8   Add the sliced onion, shrimp powder, and sesame seeds to the prepared kimchi seasoning and mix well.

9   Add the kimchi seasoning to the drained lettuce and crown daisy, and gently mix.

여름

SUMMER

김치

KIMCHI

11

오이소박이

Oi-Sobagi
Stuffed Cucumber Kimchi

오이소박이는 나의 소울 푸드 중 하나다. 햇감자가 나올 무렵에 담가 익힌 오이소박이를 찐 감자와 함께 먹는 것을 좋아한다. 소금을 넣고 감자를 찌다가 거의 다 익을 즈음 냄비 안에서 굴리면 감자에 분이 뽀얗게 앉는다. 뜨거운 감자를 집으면 저절로 막 쪼개지는데 그런 감자 한 조각에 오이소박이를 얹어 먹으면 꿀맛이 따로 없다. 그 맛은 생각만 해도 저절로 군침이 생겨 허기를 불러온다.

Oi-sobagi (stuffed cucumber kimchi) is one of my soul foods. When new potatoes come into season, I love to eat well-ripened stuffed cucumber kimchi with steamed potatoes. When potatoes are steamed with a bit of salt and rolled in the pot, they become coated in white starch. When you pick up a hot potato, it naturally falls apart into pieces. Putting a piece of stuffed cucumber kimchi on top of that hot potato piece is simply divine. Just thinking about it makes my mouth water and my stomach growl.

재료

오이 10개, 부추 100g, 양파
1/2개, 실파 5뿌리, 10%
소금물 2L(물 1.8L, 소금
200g)

김칫소 양념: 고춧가루 1컵,
멸치 액젓 1큰술, 새우젓
2큰술, 배즙 1/2컵, 다진
마늘 1큰술, 생강술 2큰술

만드는 법

1. 오이는 씻어 위와 아래쪽을 아주 조금만 잘라내고 이등분한 후 열십
자로 칼집을 넣는다.

2. 10% 소금물을 끓여 손질한 오이에 부은 다음 30분 후 건져 맑은 물
에서 씻어 물기를 뺀다.

3. 부추는 4cm 길이로 썬다.

4. 양파는 껍질을 벗기고 씻어 부추의 길이에 맞춰 곱게 채 썬다.

5. 실파는 다듬어 씻어 부추와 같은 길이로 썬다.

6. 김칫소 양념을 모두 그릇에 넣고 섞는다.

7. 준비한 김칫소 양념에 부추, 양파, 쪽파를 넣고 고루 버무려 김칫소를
만든다.

8. 손질해 둔 오이의 칼집 사이에 김칫소를 넣는다.

9. 완성된 오이소박이를 김치통에 차곡차곡 담는다.

10. 김칫소를 담았던 그릇에 물 1컵을 넣고 그릇에 묻은 양념을 모두 쓸
어 오이소박이 위에 부어 오이가 국물에 잠기게 한다.

11. 오이소박이를 꼭꼭 눌러 공기를 뺀 후 상온에서 하루나 이틀 숙성시
킨 후 냉장고에 넣고 익히면서 시원하게 먹는다.

Ingredients

Cucumbers 10, chives
100g, onion 1/2, thin
spring onions 5, 10% salt
water 2L (water 1.8L, salt
200g)

Stuffing seasoning: red
pepper powder 1 cup,
salted anchovy sauce 1
tablespoon, salted shrimp
sauce 2 tablespoons, pear
juice 1/2 cup, minced
garlic 1 tablespoon, ginger

Instructions

1. Wash the cucumbers, trim off a small portion of the ends,
then cut them in half. Make cross-shaped cuts on each
piece.

2. Boil the 10% salt water and pour it over the cucumbers.
After 30 minutes, remove the cucumbers, rinse them in
clean water and drain.

3. Cut the chives into 4cm lengths.

4. Peel, wash, and finely slice the onion to the same length
as the chives.

5. Trim, wash, and cut the thin spring onions to the same
length as the chives.

liquor 2 tablespoons

6 Put all the stuffing seasoning ingredients in a bowl and mix well.
7 Add the chives, onion, and thin spring onions to the stuffing seasoning and mix thoroughly to make the kimchi stuffing.
8 Stuff the kimchi stuffing into the cuts in the cucumbers.
9 Carefully place the stuffed cucumbers in a kimchi container.
10 Add 1 cup of water to the bowl used for mixing and scrape off any remaining seasoning. Pour this mixture over the stuffed cucumber kimchi in the container, ensuring they are submerged in the liquid.
11 Press the stuffed cucumber kimchi down firmly to remove air and let it ferment at room temperature for 1-2 days. Store it in the refrigerator and enjoy it cold while it ripens further.

12

가지소박이

Gaji-Sobagi
Stuffed Eggplant Kimchi

가지는 주로 익혀서 먹는 나물이나 냉국으로 먹지만 김치를 담가도 좋다. 살짝 찐 가지를 쭉쭉 찢어 부추와 함께 버무려 막 먹는 김치도 있지만, 오이처럼 소박이로 담그는 김치가 있다. 아삭하게 씹는 맛이 좋은 오이소박이와는 달리 부드럽고 쫄깃하게 씹히는 맛이 부추와 어우러져 제법 괜찮다.

  가지소박이를 담글 때는 중간 부분이 너무 불룩하게 큰 가지보다는 위아래가 쪽 고른 가지를 고르는 것이 좋다.

Eggplants are mainly eaten cooked, either as a vegetable side dish or chilled soups, but they are also good for making kimchi. There are different styles of eggplant kimchi, including one where slightly steamed eggplants are shredded and mixed with chives and seasoning and served immediately. Another style is stuffed eggplant kimchi (*gaji-sobagi*) that is similar to stuffed cucumber kimchi. Unlike the crunchy texture of stuffed cucumber kimchi (*oi-sobagi*), stuffed eggplant kimchi has a soft and chewy texture that complements the chives, offering a unique culinary experience.

It is best to choose evenly sized stuffed eggplant rather than those that are too bulging in the middle for making stuffed eggplant this kimchi.

재료

가지 10개, 10% 소금물
5L(물 4.5L, 소금 500g), 부추
250g, 실파 50g, 홍고추 2개

김치 양념: 고춧가루 100g,
생강즙 1작은술, 다진 마늘
1큰술, 배즙 1컵, 양파즙 1컵,
멸치 액젓 5큰술, 새우젓
3큰술

만드는 법

1 가지는 깨끗이 씻어 3cm 길이로 잘라 열십자로 칼집을 넣는다.

2 10% 소금물을 만들어 끓인다.

3 끓인 소금물을 칼집 넣은 가지에 붓고 10분간 둔다.

4 가지가 절여지면 찬물에 씻어 건져 손으로 하나씩 쥐고 살짝 물기를 제거한다.

5 부추와 실파는 깨끗하게 씻어 5mm 길이로 썬다.

6 홍고추는 반으로 갈라 씨를 제거하고 곱게 채 썬다.

7 고춧가루에 멸치 액젓과 새우젓, 다진 마늘, 생강즙, 양파즙, 배즙을 넣고 잘 섞는다.

8 7의 양념에 부추, 실파, 홍고추를 넣고 잘 버무린다.

9 8의 양념이 완성되면 식혀 놓은 가지를 넣고 전체적으로 양념에 버무린 후 칼집 사이에도 채소 양념이 들어가게 넣는다.

10 김치통에 완성된 가지김치를 양념과 함께 켜켜이 담고 상온에서 1~2일 숙성시킨 후 냉장고에 넣어두고 먹는다.

Ingredients

Eggplants 10, 10% salt
water 5L (water 4.5L, salt
500g), chives 250g, thin
spring onions 50g, red
peppers 2

Kimchi seasoning:
red pepper powder
100g, ginger juice 1
teaspoon, minced garlic
1 tablespoon, pear juice
1 cup, onion juice 1 cup,
salted anchovy sauce 5
tablespoons, salted shrimp
sauce 3 tablespoons

Instructions

1 Wash the eggplants, cut them into 3cm lengths, and make cross-shaped slits.

2 Make 10% salt water and boil.

3 Pour the boiled salt water over the cut eggplants and leave them for 10 minutes.

4 When the eggplants are salted, rinse them in cold water, drain, and gently squeeze the excess water from each eggplant using your hands.

5 Wash the chives and thin spring onions thoroughly and cut them into 5mm lengths.

6 Cut the red peppers in half lengthwise, remove the seeds, and finely shred them.

7 Mix the red pepper powder with salted anchovy sauce,

salted shrimp sauce, minced garlic, ginger juice, onion juice, and pear juice thoroughly.

8   Add the chives, thin spring onions, and red peppers to the seasoning mixture from Step 7 and mix well.

9   When the seasoning from Step 8 is done, add the cooled eggplants and coat them evenly with the seasoning. Be sure to fill the slits with the vegetable seasoning.

10 Layer the eggplant sobagi and seasoning in a kimchi container. Allow it to ferment at room temperature for 1-2 days, then store it in the refrigerator and serve as desired.

13

부추김치

Buchu-Kimchi
Chive Kimchi

부추는 마술사 같다. 한 뼘 길이의 초벌 부추를 뜯어다 간장, 고춧가루에 대충 버무려 밥상에 올리면 꼭 밥을 비비게 만든다. 초벌 부추든 재벌 부추든 혼자서도 밥반찬으로 든든한 나물이 되었다가, 오이나 가지를 만나면 부추 특유의 향과 맛으로 소박이 맛을 한껏 올려주는 마술을 부린다.

　여름 더위가 지겨워도 부추가 들어간 맛있는 김치가 있어 살 만하다는 생각이 들 정도다. 온전히 부추만으로 담근 김치도 크게 한몫한다. 큰 그릇에 수북하게 담아놓으면 반나절도 안 되어 폭 가라앉아 작은 통에 담아도 될 만큼 줄어들지만 생김치와는 다르게 익혀서 먹는 맛은 또 그대로 좋다.

Chives are like magicians. When you pick some half-grown chives that are the first to sprout that season and roughly mix them with soy sauce and red pepper powder, they make a dish so delicious that you will want to mix it with rice. Whether young or mature, chives can be a substantial side dish on their own. They will elevate the flavor of stuffed cucumber kimchi or stuffed eggplant kimchi with their unique aroma and taste.

Even when weary from the summer heat, life seems worth living when you can eat kimchi with chives. Kimchi made entirely of chives plays a significant role in this. Although a large bowl full of chive kimchi will shrink down enough to fit into a small container in less than half a day, the taste of fermented chive kimchi is just as good as fresh kimchi.

재료

부추 500g

김치 양념: 밀가루 풀 1컵(물 1컵, 밀가루 1큰술), 멸치 액젓 4큰술, 고춧가루 2/3컵, 통깨 2큰술, 생강즙 1큰술

만드는 법

1 부추는 다듬어서 깨끗하게 씻는다.

2 씻은 부추는 길이로 반을 자른다.

3 밀가루를 물에 잘 풀어 불에 올리고 저으면서 풀을 쑤어 식힌다.

4 부추의 뿌리 쪽을 아래, 잎 쪽을 위에 담고 멸치 액젓을 부어 절인다 (30분 정도).

5 부추가 절여지면서 나온 국물에 김치 양념을 넣고 잘 섞는다.

6 양념에 부추를 버무려 김치통에 담는다.

7 겉절이로 먹으면서 1~2일 상온에서 숙성시킨 후 냉장고에 넣어두고 먹는다.

Ingredients

Chives 500g

Kimchi seasoning: flour paste 1 cup (water 1 cup, flour 1 tablespoon), salted anchovy sauce 4 tablespoons, red pepper powder 2/3 cup, roasted sesame seeds 2 tablespoons, ginger juice 1 tablespoon

Instructions

1 Trim and wash the chives.

2 Cut the chives in half.

3 Dissolve the flour in water, cook while stirring constantly to make a paste, and let it cool.

4 Place the chives in a container with the root ends at the bottom and the leaf ends at the top. Pour the salted anchovy sauce over the chives and let them sit for about 30 minutes.

5 Thoroughly mix the seasonings with the liquid remaining from salting the chives.

6 Toss the chives in the seasoning mixture and place them in a kimchi container.

7 Enjoy it as fresh kimchi, ferment it at room temperature for 1-2 days, and then store it in the refrigerator.

14

오이소박이물김치

Oi-Sobagi Mul-Kimchi

Stuffed Cucumber Water Kimchi

오래전 서울에서 살 때 다니던, 오이소박이물김치를 아주 잘하는 식당이 있었다. 처음에는 밑반찬으로 내놓는데, 더 먹기를 원하면 한 사발에 1,000원을 내야 했다. 기꺼이 추가 비용을 지불하고 먹을 만한 김치였다. 아마도 우리나라에서 최초로 밑반찬을 돈 받고 판 식당이 아닐까 생각하는데, 이 물김치는 그 맛을 기억해 담근 김치다.

When I lived in Seoul long ago, I used to go to a restaurant that made the best stuffed cucumber water kimchi (*oi-sobagi mul-kimchi*). It was served as a complimentary side dish at first, but if you wanted more, you had to pay 1,000 Korean won. It was worth the extra cost. I believe it was one of the first restaurants in Korea to charge for kimchi. I make this stuffed cucumber water kimchi while reminiscing about the taste of that restaurant's kimchi.

재료

오이 10개, 부추 100g, 양파 1/2개, 실파 5뿌리, 10% 소금물 2L(물 1.8L, 소금 200g)

김칫소 양념: 멸치 액젓 1큰술, 새우젓 1.5큰술, 배즙 1/2컵, 다진 마늘 1/2큰술, 생강술 1큰술

김치 국물: 밀가루 풀 1컵(물 1컵, 밀가루 1큰술), 물 500ml, 소금 15g

만드는 법

1 오이는 씻어 꼭지와 아래쪽을 조금 잘라내고 이등분하여 열십자로 칼집을 넣는다.

2 손질한 오이에 10% 소금물을 끓여 부은 다음 식으면 건져 맑은 물에서 한 번 씻어 물기를 뺀다(30분가량 절임).

3 부추는 2cm 길이로 썬다.

4 양파는 껍질을 벗기고 씻어 부추의 길이에 맞춰 곱게 채 썬다.

5 실파는 다듬어 씻어 부추와 같은 길이로 썬다.

6 김칫소 양념 재료를 모두 넣고 버무린다.

7 준비한 양념에 부추, 양파, 실파를 넣고 고루 버무려 김칫소를 만든다.

8 손질해 둔 오이의 칼집 사이에 김칫소를 넣는다.

9 완성된 오이소박이를 김치통에 차곡차곡 담는다.

10 김칫소를 담았던 그릇에 김치 국물 재료를 넣고 그릇에 묻은 양념을 모두 쓸어 담아 오이소박이 위에 부어 오이가 국물에 잠기게 한다. (개인의 취향에 따라 국물을 더 잡아도 되고 고춧가루로 붉은 물을 들여 부어도 된다.)

11 오이소박이를 꼭꼭 눌러 공기를 뺀 후 상온에서 하루나 이틀 숙성시킨 후 냉장고에 넣고 시원하게 먹는다.

Ingredients

Cucumbers 10, chives 100g, onion 1/2, thin spring onions 5, 10% salt water 2L (water 1.8L, salt 200g)

Stuffing seasoning: salted anchovy sauce 1 tablespoon, salted shrimp sauce 1.5 tablespoons, pear juice 1/2 cup, minced

Instructions

1 Wash the cucumbers, cut off the ends slightly, cut them in half, and make slits in the shape of a cross.

2 Boil the 10% salt water and pour it over the cucumbers. Soak them in the salted water for about 30 minutes. Once cooled, remove the cucumbers, rinse them in clean water, and drain.

3 Cut the chives into 2cm lengths.

4 Peel and wash the onion, then finely slice it to match the length of the chives.

garlic 1/2 tablespoons,
ginger liquor 1 tablespoon

Kimchi liquid: flour paste
1 cup (water 1 cup, flour
1 tablespoon), water
500ml, salt 15g

5 Trim and wash the thin spring onions, then cut them to the same length as the chives.

6 Mix all the stuffing seasoning ingredients in a bowl and mix well.

7 Add the chives, onion, and thin spring onions to the stuffing seasoning and mix thoroughly to make the kimchi stuffing.

8 Stuff the kimchi stuffing into the slits in the cucumbers.

9 Carefully place the stuffed cucumbers in a kimchi container.

10 Add the kimchi liquid ingredients to the bowl used for mixing and scrape off the remaining seasoning. Pour this mixture over the stuffed cucumbers in the container, ensuring they are submerged in the liquid. (You can add more kimchi liquid or the reddish kimchi liquid colored with red pepper powder according to your preference.)

11 Press down on the stuffed cucumber firmly to remove air and let it ferment at room temperature for 1-2 days. Then store it in the refrigerator and enjoy it cold.

15

깻잎간장김치

Kkaennip Ganjang Kimchi

Perilla Leaf Soy Sauce Kimchi

우리가 고소한 향을 즐기며 쌈이나 나물로 먹는 깻잎은 참깨가 아닌 들깨의 잎이다. 들깨는 꽃이 피었다가 꽃송이가 씨앗을 품고 자라기 시작하면 잎이 노랗게 단풍이 든다. 단풍이 든 깻잎은 장아찌로 만들면 좋지만 일반적으로는 그 전에 수확한 깻잎을 조리에 이용하는 것이다.

들깨를 한 바가지 땅에 뿌리면 마치 콩나물이 자라듯 깨순이 올라온다. 6월 어느 비 오는 날 한 뼘 정도 자란 깨 모종을 옮겨 심는다. 남은 것은 뿌리를 잘라내고 나물을 해서 밥상에 올린다. 들깨의 잎이 무성하게 자라면 씨앗이 잘 앉지 않으므로 순을 쳐준다. 순을 치지 않으면 키만 클 뿐 옆으로 퍼지지 않아 수확량이 적기 때문이다. 순을 쳐서 나온 것들 중 큰 잎은 골라 김치를 담그거나 양념장을 발라 반찬으로 먹고, 작은 것들은 데쳐서 무치거나 볶아 먹는다. 그러므로 깻잎김치는 들깨가 꽃을 피우기 전에 담가 먹는 김치다. 그러나 요즘은 비닐하우스에 24시간 불을 켜놓고 들깨가 꽃을 피워 씨앗을 맺지 못하도록 하고 있어 365일 들깻잎을 구할 수 있어 좋지만 참 슬프다.

Perilla leaves, which are commonly used as wraps or in *namul* (seasoned vegetable) dishes for their nutty aroma, come from perilla plants, not sesame plants. Perilla leaves turn yellow when the flowers bloom and the seedpods start to develop. While yellowed perilla leaves are good for making pickles, we generally use the leaves harvested before they turn yellow for cooking.

When scattering a scoopful of perilla seeds on the ground, perilla sprouts emerge similar to bean sprouts. On a rainy day in June, when the sprouts are about a handspan tall, they can be transplanted. The

remaining sprouts can be trimmed at the roots and used for *namul* dishes on the meal table. As perilla plants grow thick, pruning them promotes better seed yields. Without pruning, they grow tall but not wide, resulting in a smaller harvest. The larger leaves from pruned branches are used for kimchi or side dishes, while smaller leaves are blanched and then seasoned or stir-fried. Therefore, perilla leaf kimchi is made before the perilla plants bloom. However, these days, perilla leaves are available year-round thanks to greenhouses with 24-hour artificial lighting that prevent the plants from flowering and producing seeds. This is convenient but also somewhat sad.

재료

깻잎 500g, 실고추 약간

김치 양념: 간장 1/2컵, 육수 1/2컵(2쪽 일러두기 참고), 대파 2뿌리, 다진 마늘 3큰술, 고춧가루 2큰술, 깨소금 3큰술

만드는 법

1 깻잎은 깨끗이 씻어 물기를 빼고 간추려 놓는다.
2 깻잎을 좁은 그릇에 꼭지 부분을 아래로 가지런히 담고 간장과 육수 섞은 것을 부어 30분 정도 절인다.
3 대파는 흰 부분만을 반으로 갈라 곱게 썰고 마늘은 곱게 다진다.
4 깻잎에 부어두었던 간장물만 따라 낸다.
5 간장물에 대파와 마늘, 고춧가루, 깨소금을 넣고 잘 버무린다.
6 깻잎 2장마다 한 번씩 양념을 바른 후 적당한 용기에 차곡차곡 쌓으면서 담아 꼭꼭 눌러 공기를 뺀다.
7 깻잎의 숨이 죽으면 냉장 보관해 두고 먹는다.

Ingredients

Perilla leaves 500g, red pepper threads (a small amount)

Kimchi seasoning: soy sauce 1/2 cup, broth 1/2 cup, large spring

Instructions

1 Wash the perilla leaves, drain, and set them aside neatly.
2 In a narrow container, place the perilla leaves with their stems facing downwards. Pour a mixture of soy sauce and broth over them. Allow them to soak for about 30 minutes.
3 Cut the white part of the large spring onions in half

onions 2, minced garlic 3 tablespoons, red pepper powder 2 tablespoons, roasted sesame seeds 3 tablespoons

lengthwise and chop finely. Mince the garlic.

4   Drain the soy sauce mixture from the perilla leaves.

5   Blend the soy sauce mixture with large spring onions, garlic, red pepper powder, and roasted sesame seeds and mix well.

6   Spread the seasoning mixture on every two perilla leaves and stack them neatly in a suitable container, pressing firmly to remove air.

7   Once the perilla leaves have softened, store them in the refrigerator and consume.

16

보리열무김치

Bori Yeolmu-Kimchi

Barley-Young Summer
Radish Kimchi

열무김치는 대표적인 여름 김치다. 냉장고가 없던, 내 어린 시절의 여름을 추억하면 날마다 김치를 담그시던 어머니와 이모들이 떠오른다. 그 생각을 하면 냉장고가 생필품이 된 시절을 살고 있는 사람인 나는 불평하지 말고 김치를 담가야지 하는 다짐을 하게 된다. 비가 오지 않아 물이 부족한 환경에서 자란 열무는 키가 작아 어리고 연하게 보이지만 김치를 담가 보면 억세고 질겨 먹기 불편하다. 따라서 물을 많이 먹고 자라 연한 열무를 구입해 김치를 담그는 게 좋은데 이때도 마구 주무르면 안 된다. 손이 가면 갈수록 열무가 시퍼렇게 변하면서 풋내(풀 비린내)가 나기 때문이다.

열무나 얼갈이배추로 김치를 담글 때는 조심조심 다듬고 썻고 버무려야 하는 것 외에도 풋내를 막기 위해 풀을 쑤어 넣는다. 밀가루 풀을 많이 쓰는데 열무는 보리하고도 궁합이 좋아 나는 보리밥을 지어 풀 대신 사용한다. 특히 단맛이 없는 재료들은 탄수화물을 넣어야 유산균의 먹이가 되어 새콤하고 맛있는 김치가 되는데 열무가 거기에 해당한다. 넉넉히 넣은 보리밥이 열무의 풋내도 잡고 유산균의 먹이가 되어 김치가 맛있게 익도록 돕는다. 물론 생김치로 먹어도 맛있는 열무김치다.

멸치 액젓을 넣지 않고 새우젓만으로 담가도 된다. 소금으로만 간을 해 담그면 열무의 특성이 드러나는 깔끔한 맛의 김치가 된다.

Young summer radish kimchi (*yeolmu*-kimchi) is a classic summer kimchi. The summer days of my childhood, before there were refrigerators, remind me of my mother and aunts making kimchi every day. Those memories make me realize that I, who live in a time when a refrigerator is a daily necessity, should make kimchi without complaint. Young summer radishes grown in insufficient water

conditions due to a lack of precipitation may look young and tender due to their small size, but they become tough and chewy when made into kimchi. It's better to buy young summer radishes grown with plenty of water and handle them gently to avoid turning them deep green and developing a grassy smell.

In addition to carefully washing the young summer radishes and the summer *baechu*, you can make a paste to prevent a grassy smell. While it is common to make flour paste, I use barley rice instead as it complements young summer radish well. In particular, ingredients lacking sweetness need carbohydrates to feed the lactic acid bacteria, making delicious and tangy kimchi, and young summer radish is one of them. The generous amount of barley rice eliminates the grassy smell of young summer radish and feeds the lactic acid bacteria, resulting in delicious kimchi. Of course, freshly made kimchi is also tasty.

You can make kimchi with only salted shrimp sauce, omitting the salted anchovy sauce. Seasoning solely with salt creates a clean-tasting kimchi that highlights the unique characteristics of young summer radish.

재료

열무 2kg, 10% 소금물 2L(물 1.8L, 소금 200g)

김치 양념: 고춧가루 1컵, 보리밥(보리쌀 1컵, 물 3컵), 새우젓 1/2컵, 멸치 액젓 1/2컵, 실파 100g, 다진

만드는 법

1   열무는 다듬어서 먹기 좋은 크기로 자른다.
2   손질한 열무에 10% 소금물을 부어 1시간 정도 절인다.
3   보리쌀 1컵을 씻어 3컵의 물을 붓고 푹 퍼지게 밥을 해서 식힌다.
4   실파는 다듬어 씻어 4~5cm 길이로 썰고, 청홍고추는 곱게 어슷썰기를 한다.
5   보리밥에 손질한 실파와 청홍고추, 나머지 김치 양념 재료를 잘 섞는다.

마늘 2큰술, 생강즙 1작은술,
양파즙 1/2컵, 청홍고추 각
2개

6 절인 열무를 받아놓은 물에서 아기 다루듯 살살 흔들어 씻어 건져 물
   기를 뺀다.

7 넓은 그릇에 5의 양념을 넓게 편다. 물기를 뺀 열무를 넣고 양념을 바
   르듯 살살 버무려 용기에 담는다.

8 김치를 버무린 그릇에 물과 고춧가루를 조금 넣고 소금으로 간을 맞
   춘 다음 열무김치 위에 자박하게 붓는다.

9 1~2일 상온에서 숙성시켜 냉장 보관해 두고 먹는다.

Ingredients

Young summer radish
2kg, 10% salt water 2L
(water 1.8L, salt 200g)

Kimchi seasoning: red
pepper powder 1 cup,
barley rice (barley 1 cup,
water 3 cups), salted
shrimp sauce 1/2 cup,
salted anchovy sauce 1/2
cup, thin spring onions
100g, minced garlic 2
tablespoons, ginger juice
1 teaspoon, onion juice
1/2 cup, green pepper and
red pepper 2 each

Instructions

1 Trim the young summer radishes and cut them into bite-
   sized pieces.

2 Soak the prepared young summer radishes in 10% salt
   water for about 1 hour.

3 Rinse 1 cup of barley and add the 3 cups of water. Cook
   until the barley is fully cooked and let it cool.

4 Trim and wash the thin spring onions, cut them into
   4-5cm lengths. Finely slice the green and red peppers
   diagonally.

5 Mix the cooked barley rice with the thin spring onions,
   green peppers, red peppers and other seasoning
   ingredients well.

6 Gently rinse the salted young summer radishes in the
   reserved water, as if handling a baby, then drain.

7 Spread the kimchi seasoning widely in a large bowl.
   Add the drained young summer radishes and gently mix
   them with the seasoning, coating them. Place them in a
   container.

8 Add a little water and a little bit of red pepper powder to
   the container used for mixing, season with salt, and pour
   it over the young summer radish kimchi until it is slightly
   submerged.

9 Ferment at room temperature for 1-2 days, then store in
   the refrigerator and consume.

17

열무물김치

Yeolmu Mul-Kimchi
Young Summer Radish
Water Kimchi

여름엔 시원한 곳에 있어도 뭔가 더 시원한 것이 없을까 궁리를 하게 된다. 그래서 얼음물을 마시거나 시원한 국물을 찾게 되는 것 같다. 그 시원한 것의 궁극에 잘 익은 열무물김치가 있음을 부정할 수 없다. 보리밥 한 그릇에 열무물김치 한 그릇 앞에 놓고 앉으면 한여름 더위쯤 잊고도 남는다. 밥반찬으로 먹다가 어떤 날엔 국수를 삶아 김치말이국수를 상에 올린다. 고기반찬이 부럽지 않다.

In the summer, even in a cool place, I always find myself craving something even cooler. That's why I end up drinking ice water or looking for cold soups. And there's no denying that well-fermented young summer radish water kimchi (*yeolmu mul*-kimchi) is the ultimate in coolness. Sitting down with a bowl of barley rice and a bowl of young summer radish water kimchi makes me forget all about the summer heat. I usually eat it as a side dish, but sometimes I enjoy it with noodles in chilled kimchi liquid (kimchi-*mari guksu*). It's just as satisfying as meat.

재료

열무 2kg, 물 1.8L, 소금 100g

김치 양념: 양파 1개, 빨간
파프리카 1개(매운 고추,
청홍고추는 선택), 실파
20뿌리, 채 썬 마늘 2큰술,
생강술 2큰술, 배즙 1컵, 감자
2개, 채수 4컵(다시마, 마른
표고, 파뿌리 등 자투리 채소를
우린 물)

김치 국물: 물 4L, 소금 60g
(고춧가루는 선택)

만드는 법

1   열무는 손질해 4cm 길이로 잘라 잎이 시퍼래지지 않도록 조심해서
    씻어 건진다.

2   물 1.8L에 소금 100g을 넣고 잘 풀어 소금물을 만들어 놓는다.

3   씻어 건진 열무에 2의 소금물을 붓고 가끔 열무를 뒤집으면서 1시간
    정도 절인다.

4   열무를 절이는 동안 감자의 껍질을 벗기고 잘게 썰어 채수로 삶아 으
    깬다.

5   양파는 다듬어 씻어 곱게 채 썬다.

6   파프리카는 꼭지와 씨를 제거하고 열무 크기로 썰어놓는다.

7   실파는 다듬어 씻어 4cm 길이로 자른다.

8   물 4L에 소금 60g을 넣고 잘 풀어 1.5% 소금물을 만든다.

9   준비한 재료를 김치통에 모두 넣은 후 8의 소금물을 붓는다.

10 재료가 위로 뜨지 않도록 눌러 뚜껑을 덮은 뒤 상온에 두고 발효가
    80%정도 진행되면 냉장고에 넣어 숙성시키며 먹는다.

Ingredients

Young summer radish 2
kg, 1.8L water, salt 100g

Kimchi seasoning: onion
1, red paprika 1 (spicy
pepper, red pepper,
green pepper – optional),
thin spring onion
20, shredded garlic 2
tablespoons, ginger liquor
2 tablespoons, pear juice 1
cup, potatoes 2, vegetable
stock (kelp, dried shiitake
mushrooms, spring
onion roots or leftover
vegetables) 4 cups

Instructions

1   Trim the young summer radish and cut them into 4cm
    lengths. Wash them carefully to avoid the leaves turning
    dark green. Drain them.

2   Dissolve 100g of salt in 1.8L of water to make salt water.

3   Pour the salt water from Step 2 over the drained young
    summer radish. Turn them occasionally while salting for
    about an hour.

4   While salting the young summer radish, peel and chop
    the potatoes into small pieces. Boil them in vegetable
    stock and mash them.

5   Trim, wash, and finely slice the onion.

6   Remove the stem and seeds from the paprika and cut it
    to the same size as young summer radish.

Kimchi liquid: water
4L, salt 60g (red pepper
powder optional)

7   Trim, wash, and cut the thin spring onions into 4cm
    lengths.

8   Add 60g of salt to 4L of water, stir to make 1.5% salt
    water.

9   Place all the prepared ingredients in a kimchi container
    and pour the salt water from Step 8 over them.

10 Press down on the ingredients to prevent them from
    floating above the water. Cover with the lid and leave at
    room temperature. When the kimchi reaches 80% of the
    desired fermentation, store it in the refrigerator and enjoy
    it after it ripens further.

18

얼갈이김치

Eolgari-Kimchi
Summer *Baechu*-Kimchi

얼갈이배추는 배추의 여러 품종 중 하나를 말하는 것이 아니다. 파종을 하는 시기가 달라 미처 자라지 못한 어린 배추를 얼갈이배추라고 부르는데, 다 자란 배추보다 단맛이 적다. 그렇지만 신선하며 부드러운 얼갈이배추만의 맛이 있다. 어린 배추이므로 장기 저장하는 김치로는 맞지 않다.

얼갈이배추는 열무김치 담글 때와 같이 조심스럽게 다루고 풀을 쑤어 김치를 담근다. 풀이 배추의 풋내를 막고 유산균의 먹이가 되어 김치를 새콤달콤하게 잘 익도록 해주기 때문이다. 햇홍고추를 갈아 고춧가루와 함께 쓰면 단맛이 더 나면서 색이 고운 김치가 된다.

Summer *baechu* (*eolgaribaechu*) is not a specific variety of *baechu*, but rather refers to young *baechu* that has not fully matured due to different sowing times. It has less sweetness than fully grown *baechu*, but it retains a fresh and tender *baechu* flavor. It is not suitable for long-term storage as it is a young *baechu*.

Summer *baechu*, like young summer radish kimchi, is handled carefully and made into kimchi with flour paste. This is to prevent a grassy taste and provide food for lactic acid bacteria, promoting a sweet and tangy fermentation. Using ground undried red pepper along with red pepper powder results in kimchi with a sweeter taste and more vibrant color.

재료

얼갈이배추 2kg, 10% 소금물
2L(물 1.8L, 소금 200g)

김치 양념: 고춧가루 1컵,
밀가루 풀 2컵(물 2컵, 밀가루
2큰술), 새우젓 1컵, 실파
100g, 양파 1개, 다진 마늘
2큰술, 생강즙 1작은술,
청홍고추 각 2개

만드는 법

1   얼갈이배추는 다듬어서 먹기 좋은 크기로 자른다.

2   손질한 배추에 10% 소금물을 부어 1시간 정도 절인다.

3   밀가루를 물에 잘 풀어 불에 올리고 저으면서 풀을 쑤어 식힌다.

4   실파는 다듬어 씻어 4~5cm 길이로 썰고 청홍고추는 곱게 어슷썰기
     를 한다.

5   밀가루 풀에 양념 재료를 잘 섞어 김치 양념을 만든다.

6   절인 배추를 받아놓은 물에서 아기 다루듯 살살 흔들어 씻어 건져 물
     기를 뺀다.

7   넓은 그릇에 양념을 넓게 편다. 절여 씻은 배추를 넣고 양념을 바르듯
     살살 버무려 용기에 담는다.

8   김치를 버무린 그릇에 물과 고추를 조금 넣고 소금으로 간을 맞춘 다
     음 얼갈이김치 위에 자박하게 붓는다.

9   1~2일 상온에서 숙성시켜 냉장 보관하며 먹는다.

Ingredients

Summer *baechu* 2kg, 10%
salt water 2L (water 1.8L,
salt 200g)

Kimchi seasoning: red
pepper powder 1 cup,
flour paste 2 cup (water 2
cups, flour 2 tablespoons),
salted shrimp sauce 1 cup,
thin spring onions 100g,
onion 1, minced garlic 2
tablespoons, ginger juice
1 teaspoon, red and green
pepper 2 each

Instructions

1   Trim the summer *baechu* and cut it into bite-sized pieces.

2   Pour the 10% salt water over the *baechu* and salt it for
     about 1 hour.

3   Make a flour paste by mixing flour and water well, boil,
     and let it cool.

4   Trim the thin spring onions and cut them into 4-5cm
     lengths. Thinly slice the red and green peppers diagonally.

5   Thoroughly mix the seasoning ingredients with the flour
     paste to make the kimchi seasoning.

6   Gently rinse the salted *baechu* in the reserved water, as if
     handling a baby, and drain.

7   In a wide bowl, spread the seasoning mixture widely.
     Add the drained *baechu* and gently coat it with the
     seasoning before placing it in a container.

8   Add a little water and a little bit of red pepper powder to the bowl used for mixing the kimchi, season with salt, and pour it over the summer *baechu*-kimchi.

9   Ferment the kimchi at room temperature for 1-2 days, then store it in the refrigerator before enjoying.

19

얼갈이배추물김치

Eolgaribaechu Mul-Kimchi

Summer *Baechu* Water Kimchi

이 없으면 잇몸으로 산다고 했다. 열무물김치를 대신하는 얼갈이배추물김치가 그렇다. 비슷한 것 같으면서 조금 다른 것이, 열무가 가진 쌉싸름한 맛은 없고 배추의 단맛이 사람을 홀리기 때문이다. 찬밥을 말면 숟가락질 몇 번에 올라오는 밥이 풀어내는 단맛이 또 사람을 안달 나게 만드는 김치다. 김치말이밥에도 잘 어울린다는 말을 길게 했다.

There is a saying, "If you don't have teeth, you eat with your gums." For me, summer *baechu* water kimchi (*eolgaribaechu mul*-kimchi) is like that compared to young summer radish water kimchi (*yeolmu mul*-kimchi). They're similar but somewhat different. Summer *baechu* water kimchi doesn't have the slightly bitter taste of young summer radish, and the sweetness of *baechu* is incredibly addictive. When you put cold rice into it, the sweetness of the rice that dissolves with a few spoonfuls is another thing that makes you crave more. I gave a long-winding explanation but my main point is summer *baechu* water kimchi goes well with rice in chilled kimchi liquid (kimchi-*mari bap*).

재료

얼갈이배추 2kg, 5% 소금물(물 1.9L, 소금 100g)

김치 양념: 양파 1개, 빨간 파프리카 1개(매운 고추, 청홍고추는 선택), 실파 20뿌리, 채 썬 마늘 2큰술, 생강술 2큰술, 배즙 1컵, 감자 2개, 채수 4컵(다시마, 마른 표고, 파뿌리 등 자투리 채소를 우린 물)

김치 국물: 물 4L, 소금 60g, 고춧가루 1큰술

만드는 법

1 얼갈이배추는 손질해 4cm 길이로 잘라 잎이 시퍼래지지 않도록 조심해서 씻어 건진다.

2 물 1.8L에 소금 100g을 넣고 잘 풀어 소금물을 만들어 놓는다.

3 씻어 건진 배추에 2의 소금물을 붓고 가끔 배추를 뒤집으면서 1시간 정도 절인다.

4 배추를 절이는 동안 감자의 껍질을 벗기고 잘게 썰어 채수로 삶아 으깬다.

5 양파는 다듬어 씻어 곱게 채 썬다.

6 파프리카는 꼭지와 씨를 제거하고 열무 크기로 썬다.

7 실파는 다듬어 씻어 4cm 길이로 자른다.

8 물 4L에 소금 60g을 넣고 잘 풀어 1.5% 소금물을 만든다.

9 고춧가루를 면포에 싸서 소금물에 넣고 조물조물 주물러 고춧물을 낸다.

10 준비한 재료를 김치통에 모두 넣은 후 소금물을 붓는다.

11 재료가 위로 뜨지 않도록 눌러 뚜껑을 덮고 상온에 두고 발효가 80% 정도 진행되면 냉장고에 넣어 숙성시키며 먹는다.

Ingredients

Summer *baechu* 2kg, salt water 2L (water 1.9L, salt 100g)

Kimchi seasoning: onion 1, red paprika 1 (spicy pepper or green pepper, optional), thin spring onions 20, shredded garlic 2 tablespoons, ginger liquor 2 tablespoons, pear juice 1 cup, potatoes 2, vegetable stock (kelp, dried shiitake

Instructions

1 Trim the summer *baechu* and cut them into 4cm lengths. Wash them carefully to prevent the leaves from turning dark green. Drain well.

2 Dissolve 100g of salt in 1.8L of water to make salt water.

3 Pour the salt water from Step 2 over the drained *baechu*, turning them occasionally while salting for about an hour.

4 While the *beachu* is salting, peel and chop the potatoes into small pieces. Boil them in vegetable stock until soft, then mash them.

5 Trim and wash the onion, then finely slice it.

6 Remove the stem and seeds from the paprika and cut it into the same size as the summer *baechu*.

mushrooms, spring
onion roots or leftover
vegetables) 4 cups

Kimchi liquid: water
4L, salt 60g, red pepper
powder 1 tablespoon

7   Trim and wash the thin spring onions, cutting them into
    4cm lengths.

8   Add 60g of salt to 4L of water and stir to make 1.5% salt
    water.

9   Wrap the red pepper powder in a cotton cloth and
    squeeze it into the salt water to make red pepper water.

10 Place all the prepared ingredients in a kimchi container
    and pour the salt water over them.

11 Press down on the ingredients to prevent them from
    floating above the water. Cover with a lid and leave at
    room temperature. When it reaches 80% of the desired
    fermentation, store it in the refrigerator and enjoy after it
    ripens further.

20

토마토김치

Tomato Kimchi
Tomato Kimchi

토마토는 과일인 동시에 채소다. 학술적인 분류가 어떻든 적어도 나에게는 그렇다는 얘기다. 과일처럼 먹다가 찌개에 넣기도 하고 국을 끓이는 재료로 쓰기 때문이다. 결국은 김치까지 담그게 되니 더 그렇다. 토마토김치에 쓰는 토마토는 종류에 제한을 받지 않는다. 방울토마토나 완숙 토마토 등 뭐든 괜찮다. 생김치는 샐러드 같고 잘 익은 김치는 찌개를 끓이고 싶은 맛이다.

Tomatoes are both a fruit and a vegetable. Regardless of academic classification, at least, that is how it is for me. This is because I eat them like fruit but also use them as ingredients in stews and soups. I even make kimchi with them. There are no limits to the type of tomatoes used for making kimchi. Any kind of tomato, whether cherry tomatoes or ripe tomatoes, can be used. Fresh kimchi is like a salad, and well-fermented kimchi brings to mind the flavors of stews.

재료

토마토 400g, 부추 100g,
양파 1/2개

김치 양념: 멸치 액젓 2큰술,
고춧가루 2큰술, 마늘 1쪽,
통깨 1큰술

만드는 법

1 토마토는 깨끗하게 씻어 물기를 제거하고 8등분한다.

2 부추는 다듬어 씻어 물기를 제거하고 3cm 길이로 썬다.

3 양파는 껍질을 벗기고 씻어 물기를 제거하고 부추와 같은 길이로 채
   썬다.

4 마늘은 곱게 다진다.

5 멸치 액젓과 고춧가루를 섞는다.

6 손질한 토마토, 부추, 양파, 마늘을 큰 그릇에 담고 5를 넣어 버무린다.

7 통깨를 뿌려 마무리한다.

Ingredients

Tomatoes 400g, chives
100g, onion 1/2

Kimchi seasoning:
salted anchovy sauce 2
tablespoons, red pepper
powder 2 tablespoons,
garlic 1 clove, roasted
sesame seeds 1 tablespoon

Instructions

1 Wash the tomatoes, remove moisture, and cut them into 8
   pieces.

2 Trim, wash, and drain the chives, then cut them into 3cm
   lengths.

3 Peel, wash, and drain the onion, then slice it to the same
   length as the chives.

4 Finely mince the garlic.

5 Mix the red pepper powder with salted anchovy sauce.

6 Place the prepared tomatoes, chives, onion, and garlic in
   a large bowl. Add the mixture from Step 5 and mix well.

7 Finish with sprinkled sesame seeds.

토마토김치에 쓰는 토마토는 종류에 제한을 받지 않는다. 방울토마토나 완숙 토마토 등 뭐든 괜찮다. 다만 방울토마토를 쓸 때에는 반으로 자르고, 양파나 부추도 주재료인 방울토마토의 크기에 맞춰야 보기에 좋다.

There are no limits to the type of tomatoes used for Kimchi. Any type of tomato can be used for kimchi, from cherry tomatoes to fully ripe ones. When using cherry tomatoes, it looks nice to cut chives and onions into the same size as the main ingredient.

21

고구마줄기김치

Goguma Julgi Kimchi

Sweet Potato Stem Kimchi

고구마를 심어서 잎과 줄기가 자라기 시작하면 다른 풀들이 자라지 못할 정도로 땅을 뒤덮는다. 잎과 줄기가 너무 무성하게 자라면 땅속의 고구마가 크게 자라지 않으므로 어린순을 잘라 나물로 무쳐 먹는다. 또 통통하니 길게 자란 줄기의 껍질을 벗겨 볶아 먹어도 맛나다. 그런데 껍질 벗긴 고구마 줄기를 김치로 담그면 아삭아삭하고 시원한 맛이 여름의 대표 김치인 열무김치와 견줄 만하다. 이 김치는 막 담가서 생김치로 먹어도 괜찮고 잘 익혀서 먹어도 맛있다.

특히 신맛이 도드라지게 폭 익은 김치는 생선과 함께 조려서 먹는 재미가 정말 좋다. 생선보다 고구마 줄기를 먹는 맛에 밥공기가 비는 줄도 모른다.

When sweet potatoes are planted and their leaves and stems start to grow, they cover the ground so well that other weeds cannot grow. If the leaves and stems grow too thick, the sweet potatoes underground won't grow large, so the young shoots are cut and cooked as *namul* (seasoned vegetables). It is also tasty when the thick, long stems are peeled and stir-fried. However, when the peeled stems are made into kimchi, their crisp and refreshing taste is comparable to young summer radish kimchi, a representative summer kimchi. This kimchi is best enjoyed fresh right after being made or after fermenting well.

Particularly, well-fermented kimchi with a distinct sour taste pairs delightfully with braised fish. People become so engrossed in eating the sweet potato stems that they don't even notice when their rice bowls become empty, and forget the fish altogether.

재료

손질한 고구마 줄기 2kg,
부추 100g, 실파 100g, 양파
1개, 홍고추 3개, 5% 소금물
2L(물 1.9L, 소금 100g)

김치 양념: 고춧가루 1/2컵,
다진 마늘 3큰술, 생강술
2큰술, 갈치 액젓 1컵, 밀가루
풀 2컵(물 2컵, 밀가루 2큰술)

만드는 법

1 고구마 줄기는 깨끗이 씻어 5% 소금물에 30분간 절인다.

2 절인 고구마 줄기의 껍질을 벗긴다.

3 껍질 벗긴 고구마 줄기를 한 번 씻어 5~6cm 길이로 썬다.

4 부추와 실파는 깨끗이 씻어 5~6cm 길이로 썬다.

5 양파는 껍질을 벗기고 씻어 곱게 채 썬다.

6 홍고추는 길이로 반을 갈라 속을 빼내고 5~6cm 길이로 채 썬다.

7 김치를 버무릴 큰 그릇에 김치 양념 재료를 모두 넣고 섞는다.

8 7의 양념에 고구마 줄기를 제외한 나머지 채소도 같이 넣고 고루 버
무린다.

9 8에 고구마 줄기를 넣고 다시 한번 버무린다.

10 간을 본 후 김치통에 담아 국물이 새콤하게 숙성될 때까지 상온에 두
었다가 냉장고에 넣고 익혀서 먹는다.

Ingredients

Trimmed sweet potato
stems 2kg, chives 100g,
thin spring onions 100g,
onion 1, red peppers 3, 5%
salt water 2L (water 1.9L,
salt 100g)

Kimchi seasoning: red
pepper powder 1/2
cup, minced garlic 3
tablespoons, ginger juice
2 tablespoons, salted
hairtail sauce 1 cup. flour
paste 2 cups (water 2 cups,
flour 2 tablespoons)

Instructions

1 Wash the sweet potato stems and soak them in 5% salt
water for 30 minutes.

2 Peel the sweet potato stems.

3 Rinse the peeled sweet potato stems and cut them into
5-6cm lengths.

4 Wash the chives and thin spring onions, cut them into
5-6cm lengths.

5 Peel and wash the onion, then finely shred it.

6 Cut the red peppers in half lengthwise, remove the seeds,
and shred them into 5-6cm lengths.

7 In a large bowl, mix all the kimchi seasoning ingredients.

8 Add all the vegetables except the sweet potato stems to
the kimchi seasoning from Step 7 and mix well.

9 Add the sweet potato stems to the mixture from Step 8
and mix thoroughly.

10   Taste for seasoning, place the kimchi in a container and let it ferment at room temperature until the kimchi liquid becomes pleasantly tangy. Store it in the refrigerator and consume when ready.

22

풋사과연근김치

Putsagwa Yeongeun Kimchi

Unripe Apple and Lotus Root Kimchi

연꽃이 지면 연근을 캔다. 이때 캔 연근은 배 맛이 옅게 나서 자꾸 손이 간다. 수분 함유량도 많고 아삭해서 과일처럼 먹어도 좋을 맛이다. 그래서 나는 김치로 담근다. 이 무렵 막 나오기 시작한 햇사과(풋사과)와 궁합이 좋다. 껍질을 벗길 필요도 없이 껍질과 속이 구분이 안 된다. 그러니 껍질을 먹어도 입에서 거슬리는 게 없다.

김치이지만 김치 아닌 것 같고, 김치 아닌 것 같지만 김치다. 생김치로 신선하게 맛있게 먹고, 익으면 연근과 사과의 단맛이 어우러져 다시 일품의 김치가 된다.

When lotus flowers fall, lotus roots are harvested. The lotus roots harvested at this time have a subtle pear-like flavor that makes them addictive. They are also juicy and crunchy, and one can enjoy eating them like fruit. That's why I make kimchi with them. Lotus roots pair well with the newly harvested apples (unripe apples) around this time. There is no need to peel them, as the skin and flesh are indistinguishable. So, there is no unpleasant texture when eating the skin.

It is kimchi but not quite kimchi, and it is not like kimchi, but it is kimchi. The freshly made unripe apple and lotus root kimchi is enjoyable for its fresh and delicious taste, and when fermented, the sweetness of the apple and lotus root blend together, creating another exceptional kimchi.

재료

풋사과 2개, 연근 200g

김치 양념: 연근 50g,
새우젓국 2~3큰술, 마늘
1쪽, 생강술 1작은술,
고춧가루 2큰술, 쪽파 5뿌리

만드는 법

1 풋사과는 껍질째 깨끗하게 씻어 4등분한 후 속을 제거하고 2~3mm
   두께로 썬다.
2 연근은 껍질째 잘 씻어 사과와 같은 크기로 썬다.
3 쪽파는 다듬어 깨끗이 씻은 후 2~3cm 길이로 썬다.
4 마늘은 곱게 찧는다.
5 김치 양념 재료의 연근은 강판에 간다.
6 큰 그릇에 김치 양념 재료를 모두 담아 잘 섞는다.
7 썰어놓은 사과와 연근을 먹기 직전에 6의 양념에 넣고 버무려 낸다.
8 생김치로 먹어도 맛있고, 익혀서 먹어도 좋다.

Ingredients

Unripe apples 2, lotus
root 200g

Kimchi seasoning: lotus
root 50g, salted shrimp
sauce 2-3 tablespoons,
garlic 1 clove, ginger
liquor 1 teaspoon,
red pepper powder
2 tablespoons, green
onions 5

Instructions

1 Wash the unripe apples with the skin on, cut them into
   quarters, remove the core, and slice them 2-3mm thick.
2 Wash the lotus root with the skin on, then cut it into slices
   the same size as the apples.
3 Trim, wash, and cut the green onions into 2-3cm lengths.
4 Mince the garlic finely.
5 Grate 50g of lotus root of the kimchi seasoning
   ingredients.
6 Put all the seasoning ingredients in a large bowl and mix
   well.
7 Add the sliced apples and lotus roots to the prepared
   seasoning and mix right before eating.
8 It can be enjoyed either as fresh kimchi or after
   fermentation.

23

오이지(오이지무침)

Oiji (*Oijimuchim*)
Cucumber Pickled in Salt Water (Seasoned Cucumber Pickles)

오이지는 초기 김치의 원형 같은 모습으로, 김치에서는 유물 같은 존재다. 오이와 소금으로만 담그는 것으로 냉장고가 없던 시절에 유일하게 상온에 두고 꺼내 먹었던 김치다.

항간에 물 없이 담그는 오이지라 해서 올리고당이나 물엿, 식초 등을 사용해 담그는 방법도 있지만 권하고 싶지 않다. 오이의 수분을 빼기 위해 오이보다 비싼 비용을 들이지 말고 시간만을 투자하면 된다. 그저 기다리면 된다. 담근 오이지에 골마지 혹은 할아버지라고 하는 산막 효모가 생기고 오이의 색이 노랗게 변하면 익은 것이다. 소금물 안에서 오이의 수분이 빠지면서 쪼글쪼글해지면 썰어서 무쳤을 때 아작아작한 식감을 낸다.

1~2mm 두께로 썰고 물에 1~2번 씻어 꼭 짠 후 갖은양념에 무치면 밥도둑이 된다. 아니면 찬물을 부어 오이지의 짠기를 물에 뺀 후 물과 함께 떠먹는 오이지냉국도 있다. 여름 반찬으로 이만한 것도 드물다.

먹다 남은 오이지는 씻어 꾸덕꾸덕하게 말린 뒤 된장항아리에 넣었다가 장아찌로 먹는다. 장아찌는 장을 이용한 김치와 다르지 않다.

Oiji (cucumber pickled in salt water) is similar to the original form of early kimchi and is considered a relic-like being in the kimchi world. It consists only of cucumbers and salt, traditionally stored and eaten at room temperature before refrigeration existed.

Some methods use oligosaccharides, starch syrup, vinegar, and the like, to make *oiji* without water, but these are not recommended. Rather than spending more money on ingredients than the cucumbers themselves to remove moisture from the cucumbers, you can simply invest time. Just wait. When the mold or the film

yeast also known as 'grandpa' forms on the surface of the *oiji* and cucumbers turn yellow, this means they are ripe. If the cucumbers lose moisture in salt water and become wrinkled, they develop a crunchy texture when sliced and mixed.

Slice them thinly (1-2 mm), and rinse 1-2 times in water, squeeze out the excess water, and mix with various seasonings for a delicious side dish that complements rice. Alternatively, you can make *oijinaengguk* (cold pickled cucumber soup) by pouring cold water over the *oiji* to reduce saltiness. There are few summer side dishes as refreshing as this.

Leftover *oiji* can be washed and dried until chewy, then placed in a *doenjang* (soy bean paste) jar for later enjoyment as *jangajji* (pickled vegetables). *Jangajji* is akin to kimchi using soybean paste.

재료

오이 20개

절임 물: 물 3.4L, 소금 600g

무침 양념: 파·마늘·
고춧가루·깨소금·참기름
적당량(취향에 따라 설탕·식초
약간)

만드는 법

1 오이는 씻지 말고 물기만 말린다.

2 물기를 없앤 오이를 저장용 용기에 차곡차곡 담는다.

3 냄비에 절임 물 재료를 넣고 펄펄 끓인다.

4 오이에 끓인 절임 물을 붓고 돌 등으로 눌러놓는다.

5 오이가 쪼글쪼글해지고 노랗게 숙성이 되면 소금물에서 건져 냉장고로 옮긴다.

6 먹을 때 꺼내 물에 씻은 후 얇게 썬 후 찬물에 잠시 담갔다가 냉국으로 먹거나 얇게 썬 오이지를 물에 헹궈 꼭 짠 후 파, 마늘, 고춧가루, 깨소금, 참기름으로 무쳐 밑반찬을 만든다.

Ingredients

Cucumbers 20

Brine water: water 3.4L,
salt 600g

Seasoning: spring onion,
garlic, red pepper powder,
ground sesame, sesame oil
(sugar, vinegar to taste)

Instructions

1   Do not wash the cucumbers; dry them instead.
2   Arrange the cucumbers after removing the water neatly in a storage container.
3   Put the brine ingredients in a pot and boil vigorously.
4   Pour the boiled brine over the cucumbers and press them down with a weight, such as a stone.
5   Allow the cucumbers to pickle until they become wrinkled and turn yellow. Once pickled, remove them from the salt water and transfer them to the refrigerator.
6   When ready to eat, rinse the cucumbers in water, thinly slice, and briefly soak in cold water to enjoy as a cold soup. Alternatively, rinse the thinly sliced cucumbers, squeeze out the excess water, and mix with spring onion, garlic, red pepper powder, ground sesame, and sesame oil to make a side dish.

담근 오이지에 산막 효모가 생기고 오이의 색이 노랗게 변하면 익은 것이다.
소금물 안에서 오이의 수분이 빠지면서 쪼글쪼글해지면 썰어서 무쳤을 때 아작아작한 식감을 낸다.

When the film yeast forms on the surface of the *oiji* and cucumbers turn yellow, this means
they are ripe. If the cucumbers lose moisture in the salt water and become wrinkled, they
develop a crunchy texture when sliced and mixed.

24

고추소박이

Gochu Sobagi
**Stuffed Green Pepper
Kimchi**

고추의 용도는 너무 다양하기에 김치로 변신을 한대도 아주 자연스럽다. 더구나 김치에서 빼놓을 수 없는 양념이지만 주인공이 되었을 때 그야말로 빛난다. 소를 버무리는 양념에 붉은 고춧가루가 들어가서 매운맛은 적당하므로 주인공 고추는 안 매운 것을 사용해야 누구나 먹기에 좋을 것이다. 아삭이고추 혹은 오이고추라 불리는 품종을 구입하면 된다.

Peppers have so many uses that it's quite natural to transform them into kimchi. While peppers are one of the essential ingredients in kimchi seasoning, they truly shine when they become the main ingredient. Since the red pepper powder in the kimchi stuffing seasoning provides enough spicy taste, it is best to choose non-spicy green peppers like crunch peppers or cucumber peppers so that everyone can enjoy them.

재료

고추 30개

김칫소: 인삼 1뿌리, 부추 200g, 양파 1개

김칫소 양념: 멸치 액젓 1/2컵, 고춧가루 1/2컵, 양파즙 5큰술

만드는 법

1 고추는 깨끗하게 씻어 꼭지를 1cm 남기고 자른다.

2 부추와 양파는 다듬어 깨끗이 씻어 1cm 길이로 송송 썰거나 채 썬다.

3 인삼은 부추의 길이에 맞춰 썬다.

4 고추의 한쪽에만 길이로 길게 칼집을 넣는다.

5 송송 썬 부추와 양파, 인삼에 김칫소 양념을 넣고 고루 버무린다.

6 반으로 갈라놓은 고추 속에 버무린 김칫소를 넣는다.

7 김치통에 고추소박이를 차곡차곡 담고 꼭꼭 눌러 공기를 뺀 뒤 상온에 두거나 냉장고에 넣어두고 생으로 먹거나, 상온에서 익혀 냉장고에 보관하고 먹는다.

8 김치 국물이 부족하면 물에 소금을 심심하게 녹여 자작하게 부어 익힌다.

Ingredients

Green peppers 30

Kimchi stuffing: ginseng 1 root, chives 200g, onion 1

Kimchi stuffing seasoning: salted anchovy sauce 1/2 cup, red pepper powder 1/2 cup, onion juice 5 tablespoons

Instructions

1 Wash the green peppers and cut off the stalk ends, leaving 1 cm.

2 Trim and wash the chives and onion, then chop them into 1 cm pieces or shred them.

3 Shred the ginseng to match the length of the chives.

4 Make a long cut on one side of each pepper.

5 Mix the chopped chives, shredded onion, and ginseng with the kimchi stuffing seasoning until evenly combined.

6 Fill the cut peppers with mixed kimchi stuffing.

7 Place the stuffed pepper kimchi in a container neatly, press firmly to remove air, and leave at room temperature or in the refrigerator. You can eat them raw or let them ferment at room temperature before storing in the refrigerator.

8 If there is not enough kimchi liquid, make a mild brine solution and pour it over the peppers until they are just covered and let ferment.

고추소박이용 고추는 아삭이고추 또는
오이고추라 불리는 품종을 구입하면 된다.
It is best to choose non-spicy peppers like crunch peppers
or cucumber peppers for making *gochu sobagi*.

25

미니오이소박이

Mini *Oi-Sobagi*
Stuffed Mini Cucumber Kimchi

오이소박이를 좋아한다. 하지감자를 캘 무렵에 노지에서 키운 오이도 같이 수확한다. 햇감자를 삶아 뚜껑을 덮은 채 굴리면 감자분이 뽀얗게 나온다. 뜨거운 감자를 잘 익은 오이소박이와 같이 먹으면 이보다 더 좋은 궁합이 있을까 싶다.

오이의 종류에는 오이지 담글 때 많이 쓰는 백다다기, 푸른색이 진한 취청오이, 정말 가시가 돋보이는 가시오이, 피클용 미니 오이 등이 있다. 오이소박이는 어느 품종으로 담가도 다 괜찮지만 백다다기를 가장 선호하는 분위기다. 사실 쉽게 구할 수 있는 오이가 백다다기이기도 하다.

오이소박이는 한꺼번에 많이 담갔다가 묵히면서 먹을 김치는 아니다. 그래서 품종을 가리지 않고 담가도 된다. 미니 오이로 담그면 오이를 자르지 않고 온전히 하나를 다 먹는 재미가 있다.

비닐하우스에서 대량으로 길러지는 오이의 쉽게 물러지는 특성을 어떻게 최대한 막으면서 김치를 담그는가가 관건이다. 절임용 소금물을 끓여서 부어 담그면 오이소박이가 쉬 물러지지 않는다. 오이지 담그는 방식을 차용한 방법이라 할 수 있다.

I love *oi-sobagi* (stuffed cucumber kimchi). Around the time when summer potatoes are harvested, field-grown cucumbers are also harvested. When you roll the boiled potatoes with the lid on, white potato starch comes out. I wonder if there may be no better combination than hot potatoes with well-ripened *oi-sobagi*.

There are various kinds of cucumbers: *baekdadagi*, which is often used for *oiji* (cucumbers pickled in salt water); the dark green *chwicheong* cucumber; the spiky *gasi* cucumber; and mini cucumbers

for pickles. While any kind of cucumber is suitable for *sobagi*, *baekdadagi* seems to be most preferred and is easiest to find.

Oi-sobagi is not a kimchi to make large quantities at once and store for a long time. So you can use any kind of cucumber for it. Mini cucumbers are fun because you can eat one whole without the need to cut it.

The key to making *oi-sobagi* is preventing the cucumbers, which are mass-grown in greenhouses and tend to get mushy easily, from becoming too soft. If you boil the salt water for salting and pour it over the cucumbers, they will not become too soft. This method of making *oi-sobagi* is borrowed from making *oiji*.

재료

미니 오이 40개, 5% 소금물
4L(물 3.8L, 소금 200g)
부추 200g, 양파 1개, 실파
10뿌리

김칫소 양념: 고춧가루 1컵,
멸치 액젓 2큰술, 새우젓
3큰술, 배즙 1/2컵, 다진
마늘 1큰술, 생강술 2큰술

만드는 법

1. 오이는 씻어 꼭지와 아래쪽을 조금 잘라내고 길이 쪽으로 길게 잘라 열십자 칼집을 넣는다.
2. 5% 소금물을 끓여 손질한 오이에 부은 다음 한 김 식으면 오이를 건져 맑은 물에서 한 번 씻어 물기를 뺀다.
3. 부추는 송송 썬다.
4. 양파는 껍질을 벗기고 씻어 부추의 길이에 맞춰 곱게 다지듯 썬다.
5. 실파는 다듬어 씻어 부추와 같은 길이로 썬다.
6. 그릇에 김칫소 양념 재료를 모두 넣고 섞는다.
7. 준비한 김칫소 양념에 부추, 양파, 실파를 넣고 고루 버무려 김칫소를 만든다.
8. 손질해 둔 오이의 칼집 사이에 김칫소를 넣는다.
9. 완성된 오이소박이를 김치통에 차곡차곡 담는다.
10. 김칫소를 담았던 그릇에 물 1~2컵을 넣고 그릇에 묻은 양념을 모두

쓸어 담아 오이소박이 위에 부어 오이가 국물에 잠기게 한다.

11 오이소박이를 꼭꼭 눌러 공기를 뺀 후 상온에서 하루나 이틀 숙성시킨 후 냉장고에 넣고 익히면서 시원하게 먹는다.

Ingredients

Mini cucumbers 40, 5% salt water 4L (water 3.8L, salt 200g), chives 200g, onion 1, thin spring onion 10

Stuffing seasoning: red pepper powder 1 cup, salted anchovy sauce 2 tablespoons, salted shrimp sauce 3 tablespoons, pear juice 1/2 cup, minced garlic 1 tablespoon, ginger juice 2 tablespoons

Instructions

1 Wash the cucumbers, cut off the ends slightly, and make a lengthwise slit in each cucumber, cutting in a cross shape.

2 Boil the 5% salt water and pour it over the prepared cucumbers. Once it has cooled, remove the cucumbers, rinse them in clean water, and drain.

3 Chop the chives into small pieces.

4 Peel, wash, and finely chop the onion to match the length of the chives.

5 Trim, wash, and cut the thin spring onions to the same length as the chives.

6 Put all the stuffing seasoning ingredients in a bowl and mix well.

7 Add the chives, onion, and thin spring onion to the prepared stuffing seasoning and mix thoroughly to make the kimchi stuffing.

8 Fill the kimchi stuffing into the slits of the cucumbers.

9 Carefully place the stuffed cucumbers in a kimchi container.

10 Add 1-2 cups of water to the bowl used for mixing stuffing ingredients, scrape any remaining seasoning into the water, and pour it into the stuffed cucumber kimchi container, ensuring the cucumbers are submerged in the liquid.

11 Press down stuffed mini cucumber kimchi firmly to remove air. Let it ferment at room temperature for 1-2 days before storing it in the refrigerator to ripen and enjoy cool.

26

양파김치

Yangpa Kimchi
Onion Kimchi

양파김치는 이상하다. 처음엔 양파의 매운맛이 아삭한 식감에 어우러지다가 익으면 익을수록 김치찌개 맛이 나는 김치가 된다. 막 담가서 먹으면 양파 고유의 맛이 나면서 맛있는데 익혀서 먹으면 더 맛있어지니 빠질 수밖에 없는 김치가 양파김치다.

Onion kimchi is strange. At first, the onion's spiciness blends well with its crisp texture, but as it ferments, it starts to taste like kimchi stew. When it's freshly made, you can taste the onion's unique flavor, but it gets even better when it is fermented, so it's impossible not to love onion kimchi.

재료

양파 2kg, 10% 소금물 2L(물 1.8L, 소금 200g)

김칫소: 부추 200g, 실파 100g, 홍고추 2개

김치 양념: 고춧가루 1컵, 멸치 액젓 5큰술, 새우젓 5큰술, 양파즙 2컵

만드는 법

1 양파는 작은 것을 골라 껍질을 벗기고 깨끗이 씻는다.

2 양파의 윗부분에서 뿌리 쪽으로 6등분의 칼집을 길게 넣는다.

3 칼집 낸 양파를 10% 소금물에 1시간 정도 절인다.

4 부추와 실파, 홍고추를 씻어 1cm 길이로 썬다.

5 절인 양파를 씻어 물기를 뺀다.

6 김치 양념 재료를 모두 섞은 뒤 부추와 실파, 홍고추를 넣고 버무려 김칫소를 만든다.

7 양파의 칼집 사이에 김칫소를 넣은 뒤 김치통에 차곡차곡 담는다.

8 김칫소를 버무린 그릇에 양파에서 빠진 물을 넣고 고춧가루를 조금 넣어 섞은 후 김치 위에 붓는다.

9 상온에서 1~2일 익힌 후 냉장고에 넣어 숙성시켜 먹는다.

Ingredients

Onions 2kg, 10% salt water 2L (water 1.8L, salt 200g)

Kimchi stuffing: chives 200g, thin spring onion 100g, red peppers 2

Kimchi seasoning: red pepper powder 1 cup, salted anchovy sauce 5 tablespoons, salted shrimp sauce 5 tablespoons, onion juice 2 cups

Instructions

1 Choose small onions, peel, and wash them.

2 Make 6 long slits from the top of the onion towards the root, without cutting through completely.

3 Soak the slit onions in 10% salt water for about 1 hour.

4 Wash the chives, thin spring onions, and red peppers, then finely cut them into 1cm lengths.

5 Rinse the salted onions and drain them.

6 Mix all the seasoning ingredients well, add chives, thin spring onions, red pepper and mix again to make kimchi stuffing.

7 Stuff the kimchi stuffing into the slits in the onions and place them in a kimchi container.

8 Add the remaining water from the onions to the mixing bowl, add a bit of red pepper powder and mix well. Pour this mixture over the kimchi.

9 Let it ferment at room temperature for 1-2 days, then store it in the refrigerator and enjoy after it ripens.

27

양배추양파김치

Yangbaechu Yangpa Kimchi

Cabbage-Onion Kimchi

양파는 논에서 키운 것과 밭에서 키운 것이 있다. 논에서 키운 양파보다 밭에서 키운 것이 더 단단하고 더 맵고 더 달다. 양파 생산지로 알려진 전남 무안이나 경남 함양 등지에서는 대부분 논에서 이모작으로 생산하고 있다. 오래 두고 먹을 것은 '밭 양파'를 사용해야 하고, 바로 먹을 것은 '논 양파'를 써도 괜찮다.

양파가 나오기 시작할 때면 양배추도 맛있다. 두 개의 재료로 김치를 담가 먹으면 두 배로 맛있고 두 배의 영양가가 내 몸으로 온다.

Onions can be grown in rice paddies or fields. Field-grown onions are harder, spicier, and sweeter than those grown in rice paddies. In regions like Muan or Hamyang, known for their onion production, onions are often grown in rice paddies as a double-crop. Field onions are suitable for long-term storage, while rice paddy onions are recommended for immediate consumption.

When onions are in season, cabbage is also delicious. Combining these two ingredients to make kimchi will result in a dish that is twice as delicious and nutritious.

재료

양배추 1kg, 양파 1개, 부추 50g, 중파 2뿌리

절임 물: 물 2컵, 소금 3큰술

김치 양념: 고춧가루 5큰술, 양파즙 3큰술, 멸치 액젓 3큰술, 새우젓 3큰술, 다진 마늘 1큰술, 생강술 1큰술

만드는 법

1   양배추는 한입에 먹기 좋은 크기로 썬다.

2   물 2컵에 소금 3큰술을 넣고 잘 풀어 절임 물을 만든다.

3   양배추에 절임 물을 붓고 1시간가량 절인다.

4   양파는 껍질을 벗기고 깨끗이 씻어 양배추와 같은 크기로 썬다.

5   부추는 깨끗이 씻어 물기를 빼고 3cm 길이로 썬다.

6   중파는 깨끗이 씻어 길이로 반을 가른 다음 3cm 길이로 썬다.

7   절인 양배추를 찬물에 2~3번 헹궈 건져 물기를 뺀다.

8   그릇에 김치 양념 재료를 모두 넣고 고루 섞는다.

9   김치 양념에 절인 양배추와 손질한 양파, 부추, 중파를 넣고 잘 버무린다.

10 완성된 양배추양파김치를 김치통에 꼭꼭 눌러 담아놓고 익히면서 먹는다.

Ingredients

Cabbage 1kg, onion 1, chives 50g, medium-sized spring onion 2

Salting water: water 2 cups, salt 3 tablespoons

Kimchi seasoning: red pepper powder 5 tablespoons, onion juice 3 tablespoons, salted anchovy sauce 3 tablespoons, salted shrimp sauce 3 tablespoons, minced garlic 1 tablespoon, ginger liquor 1 tablespoon

Instructions

1   Cut the cabbage into bite-sized pieces.

2   Dissolve 3 tablespoons of salt in 2 cups of water.

3   Pour the salting water over the cabbage and leave it for about 1 hour.

4   Peel the onion, wash it, and cut it into pieces the same size as the cabbage.

5   Wash the chives, drain them, and cut them into 3cm lengths.

6   Wash the medium-sized spring onions, cut them in half lengthwise, then cut them into 3cm lengths.

7   Rinse the salted cabbage in cold water 2-3 times and drain.

8   Put all the kimchi seasoning ingredients and mix well.

9   Add the salted cabbage and onion, chives, medium-sized spring onions to the kimchi seasoning and toss well.

Pack cabbage-onion kimchi tightly into a kimchi container and let it ferment before eating.

가을

AUTUMN

김치

KIMCHI

28

마배깍두기

Ma Bae Kkakdugi
Diced Chinese Yam-Pear
Kimchi

햇마가 나오면 조생종 햇배도 나온다. 김치를 잘 먹지 않는 아이들을 위해서도 그렇고, 나이 드신 어른들을 위한 신선한 김치로 햇마와 햇배로 깍두기를 담그면 인기가 좋다. 배 한 개의 껍질을 깎아 무게를 단 뒤 같은 양의 마를 더해 김치를 담그는 방식이다. 이렇게 담근 깍두 기를 마 한 번, 배 한 번 번갈아 가며 먹는 맛이 재미나다.

이 조합은 시작에 불과하다. 마, 배에 무를 더해도 되고 양파를 더해도 된다. 자신이 좋 아하는 과일과 채소를 얼마든지 조화롭게 넣어가며 담글 수 있으니 아이들과 함께 담그 기 좋은 김치다. 김치를 잘 안 먹는 아이들도 자기가 담근 김치는 잘 먹는다.

익혀서 먹어도 좋지만 조금씩 담가 한 번에 다 먹어치우는 샐러드 같은 김치로 권한다.

When fresh Chinese yams are in season, early cultivated fresh pears are also available. This diced kimchi, made with fresh Chinese yam and pear, is popular with children who typically do not enjoy kimchi, as well as with elderly people who prefer fresh kimchi. The recipe involves adding equal amounts of Chinese yam and pear, after peeling and weighing them. The alternating tastes of Chinese yam and pear in this diced kimchi are interesting.

This combination is just the beginning. You can add white radish or onion to the Chinese yam and pear. Other fruits or vegetables you like can also be included, making this kimchi a great project to make with children. Kids who are reluctant to eat kimchi enjoy the kimchi they make themselves.

While diced Chinese yam and pear kimchi is delicious after it has ripened, I recommend making it in small batches and enjoying it all at once, like a salad.

재료

배 1개(250g), 마 250g, 통깨
약간

김치 양념: 새우젓국 2큰술,
마늘 1쪽, 생강술 1큰술,
고춧가루 2큰술, 쪽파 5뿌리

만드는 법

1   마는 깨끗이 씻어 껍질을 벗기고 깍둑썰기를 한다.

2   배는 깨끗이 씻어 껍질을 벗기고 깍둑썰기를 한다.

3   쪽파는 다듬어 깨끗이 씻은 후 1cm 길이로 썬다.

4   마늘은 곱게 찧는다.

5   새우젓국에 고춧가루, 쪽파, 마늘, 생강술을 넣고 잘 섞는다.

6   썰어 놓은 마와 배를 큰 볼에 함께 담고 준비한 양념을 넣고 먹기 직
    전에 버무려 통깨를 뿌려 상에 낸다.

Ingredients

Pear 1 (250g), Chinese yam
250g, roasted sesame seeds
(a small amount)

Kimchi seasoning: salted
shrimp sauce 2 tablespoons,
garlic 1 clove, ginger liquor
1 tablespoon, red pepper
powder 2 tablespoons,
green onions 5

Instructions

1   Wash and peel the Chinese yam, then dice it.

2   Wash and peel the pear, then dice it.

3   Trim, wash, and cut the green onions into 1cm lengths.

4   Mince the garlic finely.

5   Mix the salted shrimp sauce, red pepper powder, green
    onions, garlic, and ginger liquor together.

6   Place the diced Chinese yam and pear together in a large
    bowl. Add the prepared seasonings and mix right before
    eating. Sprinkle with sesame seeds before serving.

29

도라지제피김치

Doraji Jepi Kimchi
Bellflower Roots-
Sichuan Pepper Kimchi

가을에 막 캔 도라지는 손가락으로 밀면 껍질이 벗겨질 정도로 연하고 아삭하다. 그러니 껍질을 벗길 필요도 없이 쭉쭉 찢어 김치로 담그면 아삭아삭 씹히면서 쓰고 떫은맛이 풋김치로서의 풍미를 원 없이 더해 준다. 같이 넣는 절인 배추도 칼로 썰지 않고 쭉쭉 찢어 도라지와 어우러지게 한다. 제피 가루로 마무리를 하면 제피 가루를 넣기 전과 후의 김치 맛이 전혀 다른 세계로 우리를 인도한다.

고추가 우리나라에 전파되기 전 우리네 밥상에서 매운맛을 책임지던 향신료가 바로 제피다. 물론 외국에서 수입해서 먹던 후추가 있지만 김치에 넣지는 않았다. 제피의 맛은 얼큰하니 제피 고유의 향과 함께 위로 올라 차는 매운맛을 내준다. 제피는 김치뿐 아니라 나물이나 탕 등에 두루 쓰고 있다.

Freshly harvested bellflower roots in the fall are tender and crisp, making it easy to peel off the outer skin with your fingers. When hand-shredded and made into kimchi, they add a crunchy texture and a bitter, astringent taste that enhances the flavor of the fresh kimchi. The salted *baechu* is also shredded by hand to blend well with the bellflower roots. Adding Sichuan pepper powder transforms the kimchi's flavor, creating a completely different taste experience.

Before red peppers were introduced to Korea, Sichuan pepper was the primary spice providing spiciness in Korean cuisine. Although black pepper was imported then, it was not used in kimchi. Sichuan pepper offers a spicy flavor with its unique aroma and an ascending heat that sets it apart. It is used not only in kimchi but also in various dishes such as *namul* (seasoned vegetables) and *tang* (soup).

재료

절인 배추 500g, 10% 소금물
500ml(물 450ml, 소금 50g),
도라지 200g, 무 200g, 쪽파
50g, 통깨 약간

김치 양념: 찹쌀 풀 1컵(물
1컵, 찹쌀 1큰술), 새우젓
2~3큰술, 멸치 액젓
2~3큰술, 고춧가루 2/3컵,
다진 파 2큰술, 다진 마늘
1큰술, 배 농축액 1큰술, 제피
가루 1/2~1큰술

만드는 법

1 배추는 다듬어 10% 소금물에서 5~6시간 정도 절인다.

2 도라지는 껍질을 벗겨 물에 담가 아린 맛을 뺀 후 가늘게 찢는다.

3 무는 채 썬다.

4 쪽파는 깨끗하게 씻어 4cm 길이로 자른다.

5 찹쌀 풀을 쑤어 새우젓과 함께 간다. (추젓 정도의 크기는 갈지 않아도
   된다.)

6 5에 나머지 김치 양념 재료를 모두 넣고 섞는다.

7 절인 배추를 건져 두세 번 씻은 뒤 물기를 빼서 먹기 좋은 크기로 찢
   는다.

8 채 썬 무와 가늘게 찢은 도라지에 6의 김치 양념을 넣고 치대어 간이
   배도록 한다.

9 7의 배추에 무쳐둔 무와 도라지를 넣고 버무린다.

10 취향에 따라 제피 가루의 양을 조절해서 넣고 다시 버무린다.

11 통깨를 넣어 마무리한다.

Ingredients

Salted *baechu* 500g,
10% salt water 500ml
(water 450ml, salt 50g),
bellflower roots 200g,
white radish 200g, green
onions 50g, roasted
sesame seeds (a small
amount)

Kimchi seasoning:
glutinous rice paste
1 cup (water 1 cup,
glutinous rice powder 1
tablespoon), salted shrimp
sauce 2-3 tablespoons,
salted anchovy sauce

Instructions

1 Trim the *baechu* and soak it in 10% salt water for 5-6
   hours.

2 Peel the bellflower roots, soak in water to remove the
   bitterness, then thinly shred by hand.

3 Julienne the white radish.

4 Wash the green onions and cut them into 4cm lengths.

5 Make the glutinous rice paste and grind it with the salted
   shrimp sauce. (If the shrimp are the size of *chujeot**, they
   do not need to be ground.)
   *Salted shrimp sauce made of shrimp caught in the fall

6 Combine all the kimchi seasoning ingredients together.

7 Rinse the salted *baechu* 2-3 times and drain it. Shred the
   baechu into bite-sized pieces by hand.

2-3 tablespoons, red
pepper powder 2/3 cup,
chopped spring onion
2 tablespoons, minced
garlic 1 tablespoon, pear
concentrate 1 tablespoon,
Sichuan pepper powder
1/2-1 tablespoon

8   Mix the julienned white radish and the thinly
    shredded bellflower roots with the kimchi seasoning
    from Step 6, kneading the mixture until well-coated.

9   Add the seasoned white radish and bellflower roots
    to the *baechu* from Step 7 and mix thoroughly.

10 Add the Sichuan pepper powder, adjusting the
    amount according to taste, and mix again.

11 Finish by adding sesame seeds.

가을에 막 캔 도라지는 연하고 아삭하다. 쭉쭉 찢어 김치로 담그면
아삭아삭 씹히면서 쓰고 떫은맛이 풋김치로서의 풍미를 원 없이 더해 준다.
In the fall, freshly harvested bellflower roots are tender and crisp. Hand-
shredded and fermented into kimchi, they lend a crunchy texture and a
bitter, astringent edge that sharpens the flavor of fresh kimchi.

30

고들빼기김치

Godeulppaegi-Kimchi

Bitter Lettuce Kimchi

두해살이인 고들빼기는 봄에 꽃을 피워 맺은 씨앗을 받아두었다가 여름에 다시 파종을 한다. 가을까지 충실하게 자란 잎과 뿌리를 함께 김치로 담가 먹으면 쓴맛이 살짝 올라오는 맛있는 김치가 된다. 가을에 캐지 않고 남겨둔 것들이 겨우내 땅속에서 추위를 견디다 봄에 싹을 다시 올린다. 봄에 만나는 고들빼기는 주로 뿌리가 튼실하며 잎은 가을 것보다 길이가 짧고 연하다. 봄에 채취한 것들도 김치로 담가 먹고 주로 데쳐서 무쳐도 먹는다.

고들빼기는 쓴맛이 강한 식물이다. 쓴맛을 좋아하는 사람들은 1~2시간 절인 후 바로 김치를 담가 먹지만, 이 쓴맛이 부담스러운 사람들은 짧게는 하루, 길게는 일주일 정도 옅은 소금물에 담가두어 쓴맛을 뺀 후 김치를 담근다.

먹는 사람은 몰라도 김치를 담그는 사람에게는 뿌리 부분을 손질하는 것이 여간 성가신 일이 아니다. 뿌리와 잎을 연결하는 부분에 묵은 때처럼 붙어 있는 검은 물질을 긁어내고 잔뿌리를 잘라내면서 다듬는다. 뿌리가 너무 굵으면 반으로 갈라서 절이고 잎이 너무 길면 버무리기 전 길이를 반으로 잘라 김치는 담근다.

고들빼기김치는 막 담갔을 때는 쓴 생나물 같은 맛이지만 폭 익혀 먹으면 곰삭은 맛이 일품이라 많은 사람들이 좋아하는 김치다.

Bitter lettuce (*godeulppaegi*), a biennial plant, blooms in spring; the seeds are harvested sown again in summer. The leaves and roots grow robustly until autumn when they can be made into a slightly bitter yet tasty kimchi. Over the winter, those left in the ground endure the cold and sprout again in spring. Spring *godeulppaegi* has sturdier roots and shorter, tender leaves compared to autumn *godeulppaegi*. Spring-harvested *godeulppaegi* can also be made into kimchi or

namul (seasoned vegetables) after being blanched and seasoned. *Godeulppaegi* is known for its strong bitterness. For those who enjoy the bitterness, *godeulppaegi*-kimchi can be made immediately after soaking it in salt water for only 1-2 hours. Others who find the bitterness unpleasant can soak it in mild salt water anywhere from a day to a week before making *godeulppaegi*-kimchi.

Trimming the roots can be particularly tiresome; the black residue that clings like old dirt between the roots and leaves must be scraped off and the fine roots trimmed. Thick roots are halved and soaked, while long leaves are halved before mixing the kimchi.

Freshly made *godeulppaegi*-kimchi tastes bitter like raw *namul*, but develops a deep, mellow flavor through fermentation that many people enjoy.

재료

고들빼기 1kg, 쪽파 100g, 밤 5알, 6% 소금물 2L(물 1,880ml, 소금 120g)

김치 양념: 고춧가루 1컵, 찹쌀 풀 1컵(물 1컵, 찹쌀가루 1큰술), 배즙 1컵, 조청 1큰술, 다진 마늘 2큰술, 멸치 액젓 1/2컵, 통깨 2큰술

만드는 법

1 6% 소금물을 만든다.

2 고들빼기의 지저분한 잎과 잔털 등을 제거하고 깨끗이 씻어 물기를 뺀다. (이때 뿌리가 굵은 것은 반으로 가르고 먹기 좋은 크기로 자른다.)

3 고들빼기를 6% 소금물에 담가 1시간 이상 절인다. (쓴맛을 싫어하는 사람은 반나절 정도 절인다.)

4 고들빼기를 절이는 동안 찹쌀 풀을 쑨다.

5 밤은 속껍질을 깐 후 얇게 저미거나 곱게 채 썬다.

6 쪽파는 깨끗이 씻어 3~4cm 길이로 자른다.

7 김치 양념 재료를 모두 넣고 잘 섞는다.

8 물기를 제거하고 먹기 좋은 크기로 썬 고들빼기와 밤, 쪽파를 양념에 넣고 골고루 버무린 후 김치통에 담고 꼭꼭 눌러 공기를 빼낸 뒤 상온에서 1~2일 익힌 후 냉장 보관해 두고 먹는다.

Ingredients

Bitter lettuce 1kg, green onions 100g, chestnuts 5, 6% salt water 2L (water 1,880ml, salt 120g)

Kimchi seasoning: red pepper powder 1 cup, glutinous rice paste 1 cup(water 1 cup, glutinous rice powder 1 tablespoon), pear juice 1 cup, grain syrup 1 tablespoon, minced garlic 2 tablespoons, salted anchovy sauce 1/2 cup, roasted sesame seeds 2 tablespoons

Instructions

1   Prepare 6% salt water.

2   Remove the wilted leaves and fine hairs from the bitter lettuce, then wash and drain. (If the roots are thick, cut them in half and chop into appropriate sizes to eat.)

3   Soak the bitter lettuce in 6% salt water for at least an hour. (For less bitterness, soak for about half a day.)

4   While soaking the bitter lettuce, prepare the glutinous rice paste.

5   Peel the inner skins of the chestnuts, then slice thinly or shred finely.

6   Wash the green onions and cut them into 3-4cm lengths.

7   Mix all the kimchi seasoning ingredients well.

8   Add the cut and drained bitter lettuce, the chestnuts, and the green onions to the seasoning and mix thoroughly. Pack the bitter lettuce kimchi tightly into a kimchi container to prevent air from getting in. Let the kimchi ferment at room temperature for 1-2 days and store it in the refrigerator before consuming.

봄에 만나는 고들빼기는 주로 뿌리가 튼실하며 잎은
가을 것보다 길이가 짧고 연하다. 봄에 채취한 것들도
김치로 담가 먹고 주로 데쳐서 무쳐도 먹는다.

Spring *godeulppaegi* has sturdier roots and shorter, tender
leaves compared to autumn *godeulppaegi*. Spring-harvested
godeulppaegi can also be made into kimchi or *namul*
(seasoned vegetables) after being blanched and seasoned.

31

배추김치

***Baechu*-Kimchi**
Baechu-kimchi

배추김치를 담그기 위해 내가 선호하는 배추는 2kg 전후의 무게를 가진, 잎의 길이가 짧은 배추다. 속이 너무 꽉 차지 않은 헐렁한 느낌의 배추로 김치 담그는 것을 좋아한다. 담그기도 만만하고, 한 포기씩 꺼내 썰어 먹기도 좋고, 뭐니 뭐니 해도 고소하고 단맛이 좋은 김치가 되기 때문이다.

　김장을 할 때는 배추를 재배한 지역에 서리가 두세 차례 온 후에 수확한 배추를 선택해야 한다. 서리를 맞는 동안 스스로 몸에서 수분을 뺀 배추는 농밀해져 더 달아지고 고소해지기 때문이다. 그리고 그런 배추라야 저장성이 커져서 김치를 담근 후에 물러지지 않고 오래 두고 먹을 수 있다.

　김치 담그기의 반은 배추를 절이는 일이다. 절인 배춧잎을 반으로 접었을 때 부드럽게 접힐 정도가 돼야 하지만 그렇다고 너무 절여져서 짠 배추가 되어서는 안 된다. 낮은 염도의 소금물로 긴 시간 절이면 부드럽지만 짜지 않은 배추로 절여진다.

　이 배추 절이기는 김장이 아닌 시기에도 해당이 되는데, 내한성 작물이므로 기온이 서늘할 때 자란 배추로 김치를 담그는 것이 좋다. 기온이 높을 때 자란 배추는 고랭지 배추라 하더라도 단맛이 부족하고 심심한 김치가 되기 쉽다. 맛이 부족하다고 느끼면 자꾸 이것저것 넣게 되어 시원하고 맛있는 김치에서 멀어진다.

When making *baechu*-kimchi, I prefer using *baechu* that weighs about 2 kg, with leaves that are shorter in length. I find that slightly loose *baechu* is easier to handle and cut after taking it out of the container. Most importantly, it makes kimchi that's both nutty and sweet.

The best *baechu* for *gimjang* is harvested after two or three frosts in the region where it is grown. Frost helps the *baechu* naturally shed water, making it denser, sweeter, and nuttier. This *baechu* also has better storage quality, preventing the kimchi from becoming too soft and allowing for longer storage.

Salting the *baechu* is half the battle in making kimchi. The salted *baechu* should be soft enough to bend in half easily but not overly salted. Salting for a longer period with a lower salinity results in tender yet not excessively salty *baechu*.

This principle of salting applies even outside *gimjang* season. *Baechu* is a hardy crop, and those grown in cooler temperatures are best for making kimchi. Even *baechu* grown in highlands during warm seasons tend to lack sweetness and result in bland kimchi. When the flavor seems lacking, adding more ingredients may stray from the refreshing and delicious taste of kimchi.

재료

배추 4포기, 10% 소금물 4L(물 3.6L, 소금 400g), 웃소금 1컵

김칫소: 무 2개, 고춧가루 4컵, 쪽파·미나리·갓 각 100g, 대파 4뿌리, 생새우 200g, 양파 2개, 다진 마늘 1컵, 다진 생강 1큰술, 새우젓 1컵, 멸치 액젓 1컵, 육수 4컵, 배즙 2컵

만드는 법

1 배추는 겉잎을 벗겨내고 다듬어서 반으로 가른 다음 뿌리 쪽에 칼집을 넣는다.

2 배추 절이기
① 10% 소금물을 만든다.
② 소금물에 배추를 담갔다가 건져 그릇에 담고 두꺼운 줄기 쪽에 웃소금을 얹어 차곡차곡 쌓아 절인다.
③ 5~6시간 후 배추의 위아래를 바꾸어 다시 5~6시간 더 절인다.

3 무는 깨끗이 씻어 2/3는 채 썰고 1/3은 큼직하게 썬다.

4 대파는 곱게 어슷썰기 한다.

5 쪽파와 미나리, 갓은 깨끗이 씻어 물기를 뺀 후 3~4cm 길이로 썬다.

6 양파, 생새우는 육수와 함께 갈아 다진 마늘, 다진 생강과 고루 섞는다.

7 채 썬 무에 고춧가루 1컵을 넣고 버무려 물을 들인다.

8 6의 재료에 나머지 고춧가루를 넣고 잘 섞어 불리며 숙성시킨다.

9 절인 배추는 깨끗한 물에 3~4번 헹군 후 채반에 건져 1~2시간 물기를 뺀다.

10 8의 재료에 썰어놓은 대파, 쪽파, 미나리, 갓을 넣고 잘 버무려 김칫소를 만든다.

11 배추의 바깥쪽 잎부터 들춰가며 골고루 김칫소를 넣는다.

12 겉잎으로 배추를 잘 싸서 안쪽이 하늘을 보도록 통에 꼭꼭 눌러 담는다.

13 우거지로 배추 위를 덮고 소금을 뿌린 후 뚜껑을 덮는다.

14 1~2일 정도 상온에서 숙성시킨 후 원하는 숙성의 정도보다 10% 정도 덜 숙성되었을 때 김치냉장고에 넣어두고 일주일 이상 숙성시켜 먹는다.

Ingredients

Baechu 4, 10% salt water 4L (water 3.6L, salt 400g), coarse salt (gros sel) 1 cup (for sprinkling on cabbage)

Kimchi stuffing: white radishes 2, red pepper powder 4 cups, water parsley 100g, green onions 100g, mustard leaves 100g, large spring onions 4, raw shrimp 200g, onions 2, minced garlic 1 cup, minced ginger 1 tablespoon, salted shrimp sauce 1 cup, salted anchovy sauce 1

Instructions

1 Remove the outer leaves of the baechu, trim, halve, and make small cuts at the root end.

2 Salting the baechu
① Make a 10% salt water solution.
② Soak the baechu in the salt water, then remove and place in a container. Sprinkle coarse salt on the thicker stem parts and layer the baechu in a pile.
③ After 5-6 hours, flip the baechu and repeat again for another 5-6 hours.

3 Wash the white radishes. Shred 2/3 of the radishes and cut the remaining 1/3 into larger pieces.

4 Finely slice the large spring onions diagonally.

5 Wash the green onions, water parsley, and mustard leaves, drain, and cut into 3-4cm lengths.

6 Grind the onions and raw shrimp with the broth and mix

cup, broth 4 cups, pear
juice 2 cups

the mixture with minced garlic and minced ginger evenly.

7   Mix 1 cup of red pepper powder with the shredded white
     radishes to coat them.

8   Add the remaining red pepper powder to the mixture
     from Step 6 and mix well to marinate.

9   Rinse the salted *baechu* 3-4 times in clean water, then
     drain them on a wicker tray for 1-2 hours.

10   Add the sliced large spring onions, green onions, water
     parsley, and mustard leaves to the mixture from Step 8
     and mix thoroughly.

11   Starting from the outer leaves of the *baechu*, spread the
     stuffing evenly.

12   Wrap the *baechu* tightly with the outer leaf and pack
     firmly into the container, ensuring the stuffed faces are
     upwards.

13   Cover the top with the remaining outer leaves of the
     baechu, sprinkle salt on top, and close the lid.

14   Let it ferment at room temperature for 1-2 days, then
     transfer to a kimchi refrigerator when it is about 10%
     less fermented than desired. Continue fermenting in the
     refrigerator for at least a week before consuming.

김장을 할 때는 배추를 재배한 지역에 서리가 두세 차례 온 후에 수확한
배추를 선택해야 한다. 서리를 맞는 동안 스스로 몸에서 수분을 뺀 배추는
농밀해져 더 달아지고 고소해지기 때문이다.

The best *baechu* for *gimjang* is harvested after two or three frosts in the
region where it is grown. Frost helps the *baechu* naturally shed water,
making it denser, sweeter, and nuttier.

32

백김치

Baek-Kimchi
*Baechu-*Kimchi without
Red Pepper Powder

고춧가루가 들어간 김치는 저장성이 커진다. 상대적으로 고춧가루를 넣지 않는 백김치의 종류는 오래 두고 먹기 어렵다는 이야기가 된다. 색이 변하고 군내가 나는 김치를 먹기 싫다면 조금씩 담가 빨리 먹고 새로 담가 먹는 것이 좋다. 아니면 시차를 두고 조금씩 담가 냉장고에 넣어두고 한 통씩 새로 열어 먹는 재미를 느끼는 것도 좋다. 매운 것을 조심해야 하는 어린이, 노인, 환자들의 김치로 권할 만하다.

Kimchi made with red pepper powder has a longer shelf life. This means that *baechu*-kimchi varieties without red pepper powder are relatively difficult to store for long periods. To avoid discolored and stale-smelling kimchi, it is better to make small batches and eat them quickly, then prepare a new batch. Alternatively, making small batches, storing them in the refrigerator at different times allows you to enjoy the freshness of a new container each time. This kimchi is recommended for children, the elderly, and patients who need to be careful with spicy food.

재료

배추 4포기, 10% 소금물
10L(물 9L, 소금 1kg)

김치 양념: 대파 4뿌리, 쪽파
10뿌리, 사과 1개, 배 1개, 무
1/2개, 마늘 10쪽, 생강 1쪽,
삭힌 고추 4개

김치 국물: 2% 소금물 2L(물
1,960ml, 소금 40g)

만드는 법

1 배추를 길이로 반 갈라 꼭지 부분에 칼집을 한 번 넣는다.

2 10% 소금물에 1의 배추를 담가 12시간 절인다.

3 대파와 쪽파는 다듬어 씻는다.

4 마늘과 생강을 얇게 편으로 썬다.

5 사과는 깨끗이 씻어 4등분한다.

6 배는 깨끗이 씻어 껍질을 벗기고 4등분한다.

7 무는 깨끗이 씻어 손바닥만 하게 잘라 소금에 한 번 굴린다.

8 김치통에 배추와 대파, 쪽파, 마늘 편, 생강 편 등을 켜켜이 담는다.

9 8 위에 삭힌 고추, 무, 사과, 배를 넣는다.

10 2% 소금물을 붓고 재료들이 공기와 접촉하지 않게 돌 등으로 눌러놓
   는다.

11 상온에서 1~2일 숙성시킨 후 냉장고에 넣고 발효시킨다.

Ingredients

Baechu 4, 10% salt water
10L (water 9L, salt 1 kg)

Kimchi seasoning: large
spring onions 4, green
onions 10, apple 1, pear
1, white radish 1/2, garlic
10 cloves, ginger 1 piece,
fermented pepper 4

Kimchi liquid: 2% salt
water 2L (water 1,960ml,
salt 40g)

Instructions

1 Cut the *baechu* in half lengthwise and notch the stem.

2 Soak the *baechu* from Step 1 in 10% salt water for 12
   hours.

3 Trim and wash the large spring onions and green onions.

4 Slice the garlic thinly.

5 Wash the apple and cut it into quarters.

6 Wash, peel, and quarter the pear.

7 Wash the white radish, cut it into palm-sized pieces, then
   roll them in salt.

8 In a kimchi container, layer the *baechu*, large spring
   onions, green onions, garlic slices, and ginger slices.

9 Add fermented pepper, white radish, apple, and pear to
   the container from Step 8.

10 Pour 2% salt water and press down with a stone or
   something similar to prevent the ingredients from coming

into contact with the air.

11 Let it ferment at room temperature for 1-2 days, then store it in the refrigerator to continue fermentation.

33

늙은호박김치

Neulgeun Hobak-Kimchi

Fully Ripe Pumpkin Kimchi

늙은호박김치는 황해도식 김치다. 황해도 출신인 아버지가 당신의 기억에만 의존해 알려준 재료와 맛을 어머니가 흉내 내 담기 시작하셨다. 해마다 조금씩 모양새를 잡아가며 완성한 김치로 충청도 해안가의 호박김치와는 좀 다르다. 어쩌면 배추와 무를 섞어서 만드는 섞박지와 비슷한 김치다.

늙은호박김치는 들기름을 넉넉히 두르고 지져 먹는 김치였다. 어릴 때부터 늘 그렇게 먹어온 김치였다. 그러나 내가 독립을 하고 김치를 담그면서 먹어보니 아삭하게 단맛을 내는 늙은 호박의 식감과 맛을 살린 겉절이도 맛있다. 늙은 호박의 단맛이 잘 녹아든 숙성된 김치도 물론 맛있다. 돼지고기를 몇 점 넣고 끓인 김치찌개 맛은 말이 필요 없는 음식이다.

단호박으로는 어림없는 김치 맛이니 꼭 늙은 호박으로 담그기를 권한다. 늙은 호박은 껍질이 단단하고 보기보다 가벼운 것을 골라 사야 잘 익어 맛있다. 그러니 껍질 벗길 때 손을 다치지 않게 조심해야 한다.

Neulgeun Hobak-kimchi (fully ripe pumpkin kimchi) is a style of kimchi from Hwanghae-do Province. My father, who is from Hwanghae-do, described from memory the ingredients and taste to my mother, and she began making fully ripe pumpkin kimchi. Each year, this kimchi gradually took shape and became perfected, though it differs somewhat from those made along the coast of Chungcheong-do Province. It is somewhat similar to seokbakji, which is made by mixing baechu and white radish.

Fully ripe pumpkin kimchi was traditionally braised with plenty of perilla oil, which is how I enjoyed it growing up. However, after becoming independent and making kimchi myself, I discovered that *geotjeori* (fresh kimchi), which brings out the crunchy, sweet texture of fully ripe pumpkin, is also delicious. Of course, well-fermented kimchi, where the sweetness of the ripe pumpkin is well dissolved, is equally tasty. The flavor of kimchi stew with a few pieces of pork is beyond words.

Since you can never achieve the same kimchi taste with sweet pumpkin, I recommend using a fully ripe pumpkin. When selecting a fully ripe pumpkin, choose one with hard skin that feels lighter than it looks, as it is well-ripened and flavorful. Be careful not to injure your hands when peeling it.

재료

배추 1포기(2kg 전후), 10% 소금물(물 900ml, 소금 100g), 겉소금용 굵은소금 20g, 늙은 호박 1kg

김치 양념: 고춧가루 1컵, 쪽파·미나리·갓 각 500g, 대파 1뿌리, 양파 1/2개, 간 마늘 3큰술, 간 생강 1작은술, 새우젓 1/4컵, 멸치 액젓 1/4컵, 찹쌀 풀 1컵(물 1컵, 찹쌀가루 1큰술)

만드는 법

1. 배추는 겉잎을 벗겨내고 반으로 가른다.
2. 10% 소금물에 배추를 담가 10시간 정도 절인다. (배추의 줄기 쪽엔 겉소금을 뿌리고 소금물을 붓는다.)
3. 절인 배추를 3~4번 씻어 건진 후 물기를 빼고 먹기 좋은 크기로 썬다.
4. 늙은 호박은 껍질을 벗기고 3×4cm 크기로 썬다.
5. 미나리와 갓, 쪽파는 3~4cm 길이로 자른다.
6. 대파는 어슷 썰고, 양파는 채 썬다.
7. 김치 양념 재료를 모두 넣고 잘 섞는다.
8. 썰어놓은 배추와 늙은 호박에 양념을 넣고 잘 버무린다.
9. 겉절이로 바로 먹거나 상온에서 1~2일간 숙성시킨 후 냉장고에 넣고 잘 익은 다음에 찌개로 끓여 먹는다.

Ingredients

Baechu 1 (about 2kg), 10% salt water (water 900ml, salt 100g), gros sel 20g (to sprinkle on the stem side of *baechu*), fully ripe pumpkin 1kg

Kimchi seasoning: red pepper powder 1 cup, green onions 500g, water parsley 500g, mustard leaves 500g, large spring onion 1, onion 1/2, minced garlic 3 tablespoons, minced ginger 1 teaspoon, salted shrimp sauce 1/4 cup, salted anchovy sauce 1/4 cup, glutinous rice paste 1 cup (warter 1 cup, glutinous rice powder 1 tablespoon)

Instructions

1   Remove the outer leaves of the *baechu* and cut it in half.

2   Soak the *baechu* in 10% salt water for about 10 hours. (Sprinkle gros sel on the stem side of the *baechu* and pour the salt water over it.)

3   Rinse the salted *baechu* 3-4 times, drain, and cut into bite-size pieces.

4   Peel the fully ripe pumpkin and cut into 3x4cm pieces.

5   Cut the water parsley, mustard leaves, and green onions in 3-4cm pieces.

6   Slice the large spring onions diagonally. Shred the onion.

7   Combine all the kimchi seasoning ingredients and mix well.

8   Add the seasoning to the chopped *baechu* and fully ripe pumpkin and mix thoroughly.

9   Enjoy it as fresh kimchi or let it ferment at room temperature for 1-2 days, then store it in the refrigerator until it matures well. Use it to make stew.

34

갓김치

Gat-Kimchi
Mustard Leaf Kimchi

갓은 다른 김치 재료들이 가지고 있지 않은 특유의 맛 때문에 먹는다. 전국적으로 흔한 홍갓이나 청갓 외에 여수라는 특정 지역의 돌산갓이 있다. 홍갓과 청갓은 봄과 가을에 수확을 하지만, 돌산갓은 1년에 세 번을 수확한다. 언제 수확을 하든, 아니면 몇 번의 수확을 하든 모든 갓은 너무나 매력적인 알싸한 맛을 가지고 있다.

배추김치나 동치미, 깍두기 등의 다양한 김치를 담그는 김장 때는 다른 김치의 부재료로 쓰이기도 하지만, 갓은 홀로 당당한 김치 재료로 그 개성을 드러내는 김치가 된다. 갓김치를 담글 때 쪽파를 함께 넣으면 맛이 배가된다. 막 담가서 먹으면 갓의 톡 쏘는 매운 맛이 매력적이지만 6개월 이상 묵혔다가 먹는 갓묵은지는 그 어떤 김치도 흉내 내기 어려운 환상적인 맛을 가진다. 잘 삭은 멸치 액젓의 감칠맛이 녹아든 갓묵은지를 여름에 꺼내 찬물에 만 밥과 먹으면 살면서 이런 김치를 먹어서 참 다행이구나 하는 생각이 든다.

Gat (mustard leaf) is eaten because of its unique taste that other kimchi ingredients do not have. In addition to the red _gat_ (_hong gat_) and green _gat_ (_cheong gat_) commonly found nationally, there is also a specific variety called Dolsan _gat_ from the Yeosu region. While red and green mustard leaf varieties are harvested in spring and fall, Dolsan mustard leaf is harvested three times a year. Regardless of when it's harvested or how many times, all mustard leaves have an irresistibly pungent taste.

During the _gimjang_ season, when various kimchi like _baechu-kimchi_, _dongchimi_, and _kkakdugi_ are made, mustard leaf is used as a supplementary ingredient to these kimchi. However, mustard

leaf also stands out on its own as a main ingredient in kimchi, showcasing its distinct character. When making mustard leaf kimchi, mixing it with green onions enhances the flavor. If you eat freshly made mustard leaf kimchi, the tangy, spicy taste of mustard leaf is appealing, but if you let it ferment for more than six months, the *gat-mugeunji* (fermented and aged mustard leaf kimchi) develops an extraordinary taste that no other kimchi can replicate. The savory taste of well-fermented anchovy sauce blends into the aged mustard leaf kimchi and when you take it out in the summer and eat it with rice soaked in cold water, you can't help but feel thankful for being able to enjoy such a dish in your life.

재료

갓 1kg, 쪽파 500g, 생굴 200g, 밤 5개, 잣 1큰술, 배 1개

절임: 소금 1/2컵

양념: 멥쌀 풀 1컵(물 1컵, 멥쌀가루 1큰술), 멸치 액젓 1.5컵, 고춧가루 1.5컵, 다진 마늘 3큰술, 생강즙 1큰술, 통깨 1큰술, 실고추 약간

만드는 법

1 갓과 쪽파는 깨끗이 다듬고 씻어서 물기를 뺀다.

2 갓과 쪽파에 분량의 소금을 뿌려 1시간 정도 절인 후 한 번 씻어 건진다.

3 굴은 3% 소금물에 흔들어 씻어 건져 물기를 뺀다.

4 밤은 편을 썰고, 배는 껍질을 벗겨 밤과 같은 크기로 썰고 애매한 부분은 즙을 낸다.

5 실고추는 3~4cm 길이로 썬다.

6 넓은 그릇에 멥쌀 풀과 멸치 액젓, 고춧가루를 넣고 섞은 뒤 고춧가루가 불도록 잠시 둔다.

7 젓갈에 불린 고춧가루에 4의 배즙, 다진 마늘, 생강즙, 실고추, 통깨를 넣고 잘 섞는다.

8 7에 배와 잣, 밤도 넣고 버무린다.

9 절인 갓과 쪽파를 3~4가닥씩 모아 8의 양념을 발라 통에 담는 것을 반복한다.

10 배추 우거지 등으로 한 겹 덮어 한 달 이상 숙성시켜 먹는다.

Ingredients

Gat (mustard leaf) 1kg, green onions 500g, fresh oysters 200g, chestnuts 5, pine nuts 1 tablespoon, pear 1

Salting: salt 1/2 cup

Kimchi seasoning: nonglutinous rice paste 1 cup(water 1 cup, nonglutinous rice powder 1 tablespoon), salted anchovy sauce 1.5 cups, red pepper powder 1.5 cups, minced garlic 3 tablespoons, ginger juice 1 tablespoon, sesame seeds 1 tablespoon, red pepper threads (a small amount)

Instructions

1. Trim and wash the mustard leaves and spring onions, put them neatly to remove excess water.
2. Sprinkle the required amount of salt over the mustard leaves and spring onions, and let them sit for about an hour to salt. Rinse them and drain.
3. Wash the oysters by shaking them in a 3% saltwater solution, then drain the water.
4. Slice the chestnuts, peel the pear and cut it into pieces the same size as the chestnuts. Extract juice from any remaining parts of the pear.
5. Cut the red pepper threads into 3-4cm lengths.
6. In a large bowl, mix the nonglutinous rice paste and salted anchovy sauce, red pepper powder, then let them sit for a while for the red pepper powder to soak.
7. Add minced garlic, ginger juice, red pepper threads, pear juice, and sesame seeds to the soaked red pepper powder mixture and mix well.
8. Add the pear, pine nuts, and chestnuts to the mixture from Step 7 and mix again.
9. Gather 3-4 strands of the salted mustard leaves and spring onions, and coat them with the seasoning from Step 8, then place them into a container. Repeat this process.
10. Cover the top with *baechu* leaves or similar, and let it ferment for at least a month before consuming.

홍갓과 청갓은 봄과 가을에 수확을 하지만, 돌산갓은 1년에
세 번을 수확한다. 언제 수확을 하든, 아니면 몇 번의
수확을 하든 모든 갓은 너무나 매력적인 알싸한 맛을 가지고 있다.

Red and green mustard leaves are harvested in spring and fall
while, Dolsan mustard leaves are harvested three times a year.
Regardless of the season or frequency mustard leaves always
have the season or frequency, an irresistibly pungent taste.

35

단감김치

Dangam Kimchi
Sweet Persimmon Kimchi

단감에 맛이 들면 배추도 제법 단맛이 든다. 이 두 재료가 만나 김치가 되면 별 양념을 넣지 않아도 고소하고 달달하며 매콤해 밥 없이도 손이 가는 반찬이 된다. 단감은 무르지 않고 단단한 것을 골라 껍질을 벗겨 쓴다. 아직은 배추의 단맛이 부족하므로 소금이 배추의 단맛을 최대한 끌어내도록 잘 절여서 김치를 담그면 더 맛있다.

When the sweet persimmon ripens, the *baechu* also develops a sweetness. When these two ingredients come together to make kimchi, it becomes a savory, sweet, and spicy side dish that you can't resist, even without rice. Choose a firm, not overly ripe sweet persimmon and peel it before use. Since the *baechu* is still lacking in sweetness, it's best to salt it well to draw out as much sweetness as possible, making the kimchi more delicious.

재료

절인 배추 300g, 쪽파 50g,
단감(큰 것) 1개, 볶은 통깨
1큰술

김치 양념: 멸치 액젓
3~4큰술, 고춧가루
2~3큰술, 다진 마늘 1작은술

만드는 법

1. 절인 배추는 한입에 먹기 좋은 크기로 자른다.
2. 쪽파는 깨끗하게 다듬어 씻어 3~4cm 길이로 자른다.
3. 단감은 깨끗하게 씻어 껍질을 벗긴 후 배추와 비슷한 크기와 두께로 썬다.
4. 볼에 멸치 액젓을 넣고 고춧가루와 다진 마늘을 넣고 잘 섞는다.
5. 준비해 둔 배추를 김치 양념에 잘 버무린다.
6. 버무린 배추에 단감과 쪽파를 넣고 고루 섞는다.
7. 먹기 직전에 통깨를 뿌려 마무리한다.

Ingredients

Salted *baechu* 300g, green
onions 50g, large sweet
persimmon 1, roasted
sesame seeds 1 tablespoon

Kimchi seasoning:
salted anchovy sauce 3-4
tablespoons, red pepper
powder 2-3 tablespoons,
minced garlic 1 teaspoon

Instructions

1. Cut the salted *baechu* into bite-sized pieces.
2. Trim and wash the green onions, then cut them into 3-4cm lengths.
3. Wash the sweet persimmon, peel it, and slice it into pieces similar in size and thickness to the *baechu*.
4. In a bowl, mix the salted anchovy sauce, red pepper powder, and minced garlic thoroughly.
5. Combine the prepared *baechu* with the seasoning and mix well.
6. Add the sweet persimmon and green onions to the seasoned *baechu* and mix evenly.
7. Sprinkle with roasted sesame seeds just before serving.

단감은 무르지 않고 단단한 것을 골라 껍질을 벗겨 쓴다.
Choose a firm, not overly ripe sweet persimmon. Peel it before use.

36

쪽파김치

Jjokpa Kimchi
Green Onion Kimchi

쪽파는 일 년에 두 번 수확한다. 봄 쪽파는 수확하지 않고 그냥 두면 대파처럼 꽃이 피고 잎이 쪼그라들며 뿌리가 굵어진다. 굵어진 뿌리 씨앗은 여름까지 보관했다가 파종해 가을 김장철에 맞춰 수확한다.

그래서 나도 쪽파 수확에 맞춰 해마다 두세 번 김치를 담근다. 봄에는 어리고 짧은 쪽파로 담기도 하고, 굵어진 뿌리를 가진 쪽파로 김치를 담가 가을까지 묵혔다가 먹는다. 뿌리의 매운맛이 사라지고 달며 잘 익은 깊은 맛이 어떤 음식과도 잘 어울린다. 가을 김장철에 담근 김치는 겨울내 먹다가 봄이 되어 햇쪽파가 나오면 생선을 넣고 지져 먹으며 마무리한다.

쪽파를 다듬을 때는 송곳처럼 뾰족한 잎의 끝부분을 살짝 잘라내야 한다. 안 그러면 익으면서 생긴 기포들로 부풀어 오른 쪽파가 씹을 때 입안에서 빵 소리를 내면서 터지는 일이 생긴다.

어머니는 아침 밥상에 쪽파김치를 올리지는 않으셨다. 밖에 나가는 식구들의 입에서 김치 냄새 난다고 걱정하시면서. 그리고 고구마 삶으면 쪽파김치와 함께 주셨는데 고구마 먹고 불편한 속을 달래게 하셨던 것 같다.

Green onions are harvested twice a year. If spring green onions are left unharvested, they will grow flowers like large spring onions, the leaves will shrink, and the roots will become thick. The thickened root seeds are stored until summer and then sown to be harvested in time for the fall *gimjang* season.

I also make kimchi two or three times every year according to the

green onion harvest. In spring, I make kimchi with young and short green onions or with green onions that have thickened roots and store them until fall. The spicy taste of the roots disappears, and the sweet and well-ripened deep flavor goes well with any dish. The green onion kimchi made during the fall *gimjang* season is eaten throughout the winter, and when spring comes and fresh green onions are available, it is finished by braising it with fish.

When trimming green onions, you should cut off the sharp, pointed ends of the leaves. Otherwise, green onions may swell up with air bubbles during fermentation, and they might burst in your mouth with a popping sound when you eat them.

My mother did not serve green onion kimchi for breakfast. She was worried that the smell of kimchi would linger on the breath of family members going out in public. When she boiled sweet potatoes, she served them with green onion kimchi, probably to soothe the stomach after eating sweet potatoes.

재료

쪽파 800g

김치 양념: 밀가루 풀 1컵(물 1컵, 밀가루 1큰술), 고춧가루 1컵, 멸치 액젓 3큰술, 새우젓 3큰술, 통깨 2큰술, 배즙 1컵, 생강즙 1작은술

만드는 법

1 쪽파는 다듬어서 깨끗이 씻는다.
2 쪽파를 뿌리 부분이 아래쪽으로 가도록 큰 볼에 담고 멸치 액젓과 새우젓을 넣고 절인다.
3 밀가루를 물에 잘 풀어 불에 올리고 저으면서 풀을 쑤어 식힌다.
4 뿌리 쪽이 어느 정도 절여지면 잎 쪽도 뉘어 같이 절인다(1시간 정도).
5 쪽파가 절여지면서 나온 국물에 김치 양념 재료를 넣고 잘 섞는다.
6 양념에 쪽파를 살살 버무려 한입에 먹기 좋게 돌돌 말아 김치통에 담는다.
7 1~2일간 상온에서 숙성시킨 후 냉장고에 넣어두고 먹는다.

Ingredients

Green onions 800g

Kimchi seasoning: flour paste 1 cup(water 1 cup, flour 1 tablespoon), red pepper powder 1 cup, salted anchovy sauce 3 tablespoons, salted shrimp sauce 3 tablespoons, roasted sesame seeds 2 tablespoons, pear juice 1 cup, ginger juice 1 teaspoon

Instructions

1   Trim the green onions and wash them.

2   Place the green onions in a large bowl with the roots at the bottom and soak them in the salted anchovy sauce and the salted shrimp sauce.

3   Dissolve the flour in water, boil it while stirring to make the paste. Let it cool.

4   Once the root part is salted, lay down the leafy part of the green onions and marinate them together (for about an hour).

5   Add the kimchi seasoning to the liquid from the marinated green onions and mix well.

6   Gently toss the green onions in the seasoning, roll them up into bite-sized pieces, and place them in a kimchi container.

7   Ferment at room temperature for 1-2 days, then store it in the refrigerator and enjoy.

37

총각김치

Chonggak-Kimchi
Whole Radish Kimchi

총각김치는 맛있다. 그런데 맛있는 총각김치를 담그려면 맵지 않고 단맛이 있는 총각무를 잘 골라 사야 한다. 너무 크지 않아야 하고 밑이 통통하니 긴 꼬리가 있어 마치 쥐의 엉덩이 같은 모양의 무가 좋다. 잔뿌리가 많지 않은 것을 골라 구입해 꼬리와 잔털은 잘라내고 통으로 절인다. 조금 큰 무는 절여서 썻은 후에 먹기 좋은 크기로 자르는 것이 좋다.

총각김치는 80% 정도 익었을 때 냉장고에 넣어 20%를 마저 익히는 것이 정석이다. 그래야 국물만 익은, 총각무 속의 간도 맛도 덜 든 김치를 먹지 않는다. 국물과 무가 적절하고 조화롭게 익은 김치가 군내도 없고 맛있다.

Whole radish kimchi (*chonggak*-kimchi) is delicious. However, in order to make delicious whole radish kimchi, you should select radishes that are not spicy and have a sweet taste. They should be moderate in size, with a long tail and a plump bottom resembling a mouse's buttocks. Choose whole radishes with few rootlets, remove the tail and fine hairs, and salt them whole. For slightly larger radishes, it is better to cut them into pieces after salting and rinsing them.

The standard method for making whole radish kimchi is to store the kimchi in the refrigerator when it is 80% ripe and allow the remaining 20% to ripen in the refrigerator. This ensures that you will avoid consuming kimchi that is only fermented in kimchi liquid and has yet to season the inside of the radish. Kimchi that has been fermented with both kimchi liquid and radish develops a delicious flavor without any stale smell.

재료

총각무 2kg, 10% 소금물
2L(물 1.8L, 소금 200g)

김치 양념: 쪽파 300g, 대파
1뿌리, 다진 마늘 5큰술,
생강술 1/2컵, 멸치 액젓
1/3컵, 새우젓 1/3컵,
고춧가루 1컵, 배즙 1컵,
육수 1컵, 멥쌀 풀 1컵(물 1컵,
멥쌀가루 1큰술)

만드는 법

1   총각무는 잔털을 제거하고 무청과 연결된 부분의 검은 부분은 칼로
    도려낸다.

2   손질한 총각무를 10% 소금물에 담가 1~2시간 정도 절인다.

3   쪽파는 다듬어서 깨끗이 씻은 다음 3~4cm 길이로 자른다.

4   대파는 깨끗하게 다듬어 어슷하게 썬다.

5   육수를 만든다.

6   그릇에 김치 양념 재료를 모두 넣고 섞는다.

7   절인 총각무를 건져 깨끗하게 씻은 후 물기를 제거한다.

8   총각무에 김치 양념을 넣고 잘 버무린다.

9   완성된 총각김치를 김치통에 꼭꼭 눌러 담아 상온에서 2~3일 두고
    살피다가 국물이 먼저 익으면 냉장고에 넣고 일주일 이상 익혀서 먹
    는다.

Ingredients

Whole radish 2kg, 10%
salt water 2L (water
1.8L, salt 200g)

Kimchi seasoning:
green onions 300g,
large spring onion
1, minced garlic 5
tablespoons, ginger
liquor 1/2 cup, salted
anchovy sauce 1/3 cup,
salted shrimp sauce
1/3 cup, red pepper
powder 1 cup, pear
juice 1 cup, broth 1
cup, nonglutinous rice
paste 1 cup (water 1
cup, nonglutinous rice

Instructions

1   Trim the rootlets of the whole radishes and remove any
    black parts connected to the radish greens.

2   Soak the trimmed whole radishes in 2L of 10% salted
    water for 1-2 hours.

3   Trim and wash the green onions, then cut them into 3-4
    cm lengths.

4   Wash the large spring onion and slice it diagonally.

5   Prepare the broth.

6   Put all the kimchi seasoning ingredients in a bowl and
    mix well.

7   Rinse the salted whole radishes and drain.

8   Add the kimchi seasoning to the whole radishes and mix
    well.

9   Pack the whole radish kimchi tightly into a kimchi
    container, leave at room temperature for 2-3 days.

powder 1 tablespoon)

Once the kimchi liquid begins to ferment, store it in the refrigerator and allow it to ferment for more than a week before consuming.

38

총각무물김치

Chonggakmu Mul-Kimchi

Whole Radish Water Kimchi

총각무로 물김치를 담그면 약간 다른 모양의 동치미를 먹는 느낌이 난다. 실제로 총각무를 사용해 동치미를 담그는 사람들도 있다. 큰 무를 사용하는 것과는 달리 더 아삭하여 씹는 맛이 좋다. 잘 익어 시원한 국물과 무는 소화를 잘되게 돕는다. 밥을 먹다가도 소면을 삶아 국수말이를 해 먹고 싶은 생각이 들게 한다.

When you make water kimchi with whole radish (*chonggakmu*), it feels like eating a slightly different kind of *dongchimi* (white radish kimchi in water). In fact, some people use whole radish to make *dongchimi*. Unlike with large white radishes, whole radish is crisper and fine chewy texture. The well-fermented, cool broth and radish are helpful for digestion. It always makes me want to boil some noodles and make noodles in chilled whole radish water kimchi.

재료

총각무 2단, 쪽파 300g, 10%
소금물 5L(물 4.5L, 소금
500g)

김치 국물: 마늘 10쪽, 생강
1쪽, 2% 소금물 5L(물 4.9L,
소금 100g), 찹쌀 풀 3컵(물
3컵, 찹쌀가루 3큰술)

만드는 법

1 총각무는 손질하여 10% 소금물에 2시간 정도 절인다.

2 절인 총각무를 깨끗이 씻어 먹기 좋은 크기로 자른다.

3 찹쌀 풀을 쑨다.

4 마늘과 생강은 껍질을 까고 씻어 다진다.

5 쪽파는 다듬어 씻어 5cm 길이로 썬다.

6 그릇에 김치 국물 재료를 모두 넣고 고루 섞는다.

7 김치통에 총각무를 넣고 만들어둔 김치 국물을 붓는다.

8 총각무가 떠오르지 않게 눌러두고 상온에서 이틀간 숙성시킨 후 냉장
   고에 넣어두고 일주일 이상 숙성시켜 무가 삭으면 먹어도 좋다.

Ingredients

Whole radish 2
bunches, green onions
300g, 10% salt water
5L (water 4.5L, salt
500g)

Kimchi liquid: garlic
10 cloves, ginger 1
piece, 2% salt water 5L
(water 4.9L, salt 100g),
glutinous rice paste
3 cups (water 3 cups,
glutinous rice powder 3
tablespoons)

Instructions

1 Trim the whole radishes and soak them in 10% salt water
   for about 2 hours.

2 Rinse the salted whole radishes and cut them into bite-
   sized pieces.

3 Prepare the glutinous rice paste.

4 Peel, wash and mince the garlic and ginger.

5 Trim and wash the green onions, cut them into 5cm
   lengths.

6 In a bowl, put all the kimchi liquid ingredients and mix
   well.

7 Place the whole radishes in a kimchi container and pour
   the kimchi liquid over them.

8 Press down on the whole radishes to prevent them from
   floating above the water. Leave them to ferment at room
   temperature for 2 days. Store them in the refrigerator and
   eat after at least 1 week of fermentation when the radishes
   have softened.

겨울

WINTER

김치

KIMCHI

39

굴깍두기

Gul-Kkakdugi
Oyster-Diced White Radish
Kimchi

가을무를 저장해 두었다가 겨울에 굴깍두기를 담글 때는 특별히 더 맛있고 좋은 무를 찾아 다닐 필요가 없다. 이 무렵의 무는 웬만하면 다 달고 시원한 맛을 가진다. 무보다는 좋은 굴을 선택하는 것이 더 중요하다. 우리가 생굴로 먹는 대부분의 굴은 남해안 쪽에서 수하식으로 양식한 씨알이 큰 굴이라 깍두기를 담그는 데 사용하기는 애매하다. 서해에서 바닷물이 드나들 때 하루 한 번 이상 공기 중에 노출되면서 더디게 자라 씨알이 작은 굴이 김치를 담그기에는 더 적합하다.

굴깍두기는 막 담가 먹으면 굴무침 같은 느낌이다가 숙성되면 굴이 내주는 단맛이 무와 만나 그 시원함이 입안에서 터진다. 그러다 깍두기 속의 굴이 곰삭으면 어디다 비할 수 없는 감칠맛을 낸다. 국수를 말거나 밥을 비벼도 좋고, 특히 밥을 볶으면 아주 훌륭한 한 그릇 요리가 된다.

There is no need to search for special white radishes when making oyster-diced white radish kimchi (*gulk-kakdugi*) in the fall. Most white radishes at this time of year are naturally sweet and refreshing. It is more important to choose good oysters. Most oysters we eat raw are large, cultivated using the hanging culture method from the south coast, and thus are not suitable for *kkakdugi*. Smaller oysters from the west coast, exposed to air at least once a day as the tides change, are better suited for kimchi.

Freshly made oyster *kkakdugi* tastes like oyster *muchim* (seasoned oysters), but as it ferments, the sweetness of the oysters blends with the white radishes, creating an explosion of refreshing flavors. As

the oysters within the *kkakdugi* further ferment, they develop an unparalleled savory taste. It is great if mixed with noodles or rice and is especially delicious when used in fried rice, turning it into a fantastic one-bowl meal.

재료

무 2개(2kg), 굴 300g, 3% 소금물 3L(물 2,910ml, 소금 90g), 미나리 100g, 쪽파 100g, 갓 100g

깍두기 양념: 고춧가루 1컵, 대파 1뿌리, 다진 마늘 3큰술, 다진 생강 2작은술, 새우젓 2/3컵, 멸치 액젓 2/3컵, 배즙 1/2컵

만드는 법

1 무는 껍질째 깨끗이 씻어서 사방 2cm 크기로 자른다.

2 바닷물과 같은 염도인 3% 소금물을 1L씩 나눈 뒤 굴을 넣고 살살 흔들면서 3회에 걸쳐 깨끗이 씻는다.

3 미나리, 쪽파, 갓은 다듬어 씻어 2cm 길이로 썬다.

4 대파는 어슷하게 썬다.

5 새우젓은 곱게 다진다.

6 그릇에 깍두기 양념을 모두 넣고 섞는다.

7 6의 양념에 무, 미나리, 쪽파, 갓을 넣고 버무린다.

8 골고루 버무려졌으면 굴을 넣고 다시 한번 살살 버무린다.

9 항아리나 김치통에 굴깍두기를 꼭꼭 눌러 담고 상온에서 1~2일간 숙성시켜 냉장고에 넣어두고 먹는다.

10 김장철에는 상온에서 열흘 정도 숙성시키면 알맞게 익는다.

Ingredients

White radish 2 (2kg), oysters 300g, 3% salt water 3L (water 2,910ml, salt 90g), water parsley 100g, green onions 100g, mustard leaves 100g

Kimchi seasoning: red pepper powder 1 cup, large spring onion

Instructions

1 Wash the white radishes with the skin on and cut them into 2cm cubes.

2 Rinse the oysters gently in a 3% brine solution (similar to seawater salinity) 3 times.

3 Trim and wash water parsley, green onions, mustard leaves, and cut them into 2 cm lengths.

4 Slice the large spring onion diagonally.

5 Finely chop the salted shrimp sauce.

1, minced garlic 3 tablespoons, minced ginger 2 teaspoons, salted shrimp sauce 2/3 cup, salted anchovy sauce 2/3 cup, pear juice 1/2 cup

6   Combine all the seasoning ingredients in a bowl and mix well.

7   Add the white radishes, water parsley, green onions, mustard leaves to the seasoning mixture from Step 6 and toss to coat.

8   Once everything is well mixed, add the oysters and gently mix again.

9   Pack the mixture tightly into an earthenware pot or kimchi container and let it ferment at room temperature for 1-2 days before storing it in the refrigerator.

10 During the *gimjang* season, ferment the kimchi at room temperature for about 10 days to properly ripen.

굴깍두기는 막 담가 먹으면 굴무침 같은 느낌이다가 숙성되면
굴이 내주는 단맛이 무와 만나 그 시원함이 입안에서 터진다.
그러다 깍두기 속의 굴이 곰삭으면 어디다 비할 수 없는 감칠맛을 낸다.

At first, oyster *kkakdugi* tastes similar to oyster *muchim* (seasoned
oysters). As fermentation progresses, the oysters' sweetness mingles
with the white radish, resulting in a crisp, refreshing flavor. Over
time, the oysters develop a savory depth that's hard to match.

40

나박김치

Nabak-Kimchi

Sliced White Radish Water
Kimchi

어머니는 나박김치를 일 년에 딱 두 번 담그셨다. 설날과 추석 밑에 담가 잘 익혔다가 차례 상에 올리는 김치로 쓰셨다. 맛있는 배추와 무가 큰 몫을 하는 김치라 더 그랬을 것 같다. 그래서 나도 일 년에 두 번 나박김치를 담근다.

배추와 무의 단맛이 나박김치 국물의 달고 시원한 맛을 책임지도록 해야 한다. 설탕이나 인공 감미료 등에 맡길 일은 절대 아니다. 제철이 아닐 때 김치를 담그면 맛을 흉내 내기 위한 재료들을 첨가해야 하니 좋지 않은 방법이다.

나박김치의 이름은 재료들을 나박나박 썰어 담는 데서 연유했다. 가로 2cm, 세로 3cm, 높이 1~2mm 정도의 크기로 써는 방법을 나박썰기라고 한다. 숟가락으로 김치를 뜨면 밖으로 빠져나가지 않고 한입에 쉽게 먹을 수 있는 크기다.

My mother used to make sliced white radish water kimchi (*nabak-kimchi*) only twice a year. She made it before the Lunar New Year and Chuseok (Korean Thanksgiving), fermented it well, and offered it to our ancestors. I believe this is because *nabak*-kimchi relies on the delicious flavors of *baechu* and white radish. Thus, I also make *nabak*-kimchi twice a year.

The sweetness of the *baechu* and white radish should be responsible for the sweet and refreshing taste of the *nabak*-kimchi liquid. It is definitely not something to be enhanced with sugar or artificial sweeteners. It is not ideal to make kimchi out of season because you have to add ingredients to imitate the natural taste.

The name "*nabak*-kimchi" comes from the fact that the ingredients

are cut into thin slices. The method of cutting pieces 2cm wide, 3cm long, and 1-2mm thick is called *nabak* slicing. This size prevents spills when scooping the kimchi with a spoon and makes it easy to eat in one bite.

재료

무 500g, 배추속대 300g, 사과 1/2개, 소금 3큰술, 대파 1뿌리, 마늘 5쪽, 생강 1쪽, 미나리 10대, 실고추 약간

김치 국물: 고춧가루 2큰술, 배즙 1컵, 육수 1컵, 1.5% 소금물 3L(물 2,955ml, 소금 45g)

만드는 법

1 무는 깨끗이 씻어서 나박나박 썬다(3cm×2cm×0.2cm).

2 배추는 흰 속대만 골라 씻어 잎을 길이로 이등분한 다음 3cm 길이로 썬다.

3 무와 배추를 소금에 30분간 절인다.

4 대파는 2cm 길이로 잘라 곱게 채 썬다.

5 마늘과 생강도 껍질을 벗겨 깨끗이 씻은 후 곱게 채 썬다.

6 미나리는 잎을 모두 따내고 깨끗이 씻어 2cm 길이로 썬다.

7 사과는 깨끗하게 씻어 껍질째 나박나박 썬다.

8 절인 무와 배추는 헹궈서 체에 밭쳐 물기를 뺀다.

9 모든 재료를 넣고 고루 섞어 김치통에 담는다.

10 1.5% 소금물 3L에 배즙과 육수를 넣고 잘 섞는다.

11 고춧가루를 면포에 잘 싸서 10의 김치 국물에 넣고 흔들어 고추의 붉은색과 맛이 우러나게 한다.

12 9의 재료에 11의 김치 국물을 부어 익힌다.

13 국물이 익기 시작하면 낮은 온도의 냉장고나 밖에 두고 천천히 익혀 먹는다.

Ingredients

White radish 500g, *baechu* heart 300g, apple 1/2, salt 3 tablespoons, large

Instructions

1 Wash the white radish and cut it into thin slices (3cm×2cm×0.2cm).

2 Wash the *baechu* hearts (white hearts only), cut them in

spring onion 1, garlic 5 cloves, ginger 1 piece, water parsley 10 stalks, red pepper threads (a little bit)

Kimchi liquid: red pepper powder 2 tablespoons, pear juice 1 cup, broth 1 cup, 1.5% salt water 3L (water 2,955ml, salt 45g)

half lengthwise, and then cut into 3cm lengths.

3. Preserve the white radish and *baechu* hearts in salt for 30 minutes.

4. Cut the large spring onion into 2cm pieces and finely shred.

5. Peel the garlic and ginger, wash, and finely shred.

6. Remove all the leaves from the water parsley, wash and cut into 2cm lengths.

7. Wash the apple and cut it into thin slices with the skin on.

8. Rinse the salted white radish and *baechu* hearts, then drain.

9. Mix all the prepared ingredients well and place them in a kimchi container.

10. Mix pear juice and broth with 3L of 1.5% salt water.

11. Wrap the red pepper powder in a cotton cloth and shake it in the kimchi liquid from Step 10 to bring out the red pepper color and flavor.

12. Pour the kimchi liquid from Step 11 over the ingredients from Step 9 and let it ferment.

13. When the kimchi liquid starts to ferment, store it in a low-temperature refrigerator or outside and let it ferment slowly before eating.

나는 일 년에 두 번 나박김치를 담근다.
배추와 무의 단맛이 나박김치 국물의
달고 시원한 맛을 책임지도록 해야 한다.
I make *nabak*-kimchi twice a year. The
natural sweetness of the baechu and
white radish is what makes the liquid
taste light, sweet, and refreshing.

41

시금치겉절이

Sigeumchi Geotjeori
Spinach Fresh Kimchi

겨울에 푸른색을 가진 채소를 밭에서 만나기란 참으로 어렵다. 그런 겨울에 거의 유일하게 만날 수 있는 채소가 시금치다. 시금치는 내한성 작물이라 눈이 하얗게 쌓인 노지에서도 얼어 죽지 않고 살아남는다. 추울수록 자세를 낮추고 땅에 납작 엎드려 마치 그림처럼 보일 만큼 부피감이 없이 지낸다. 언 땅에서 시금치를 칼로 캐어 나물로 무치면 단맛이 줄줄 흐르는 느낌이 난다. 분홍색 뿌리까지 다듬어 겉절이로 무친다. 건조한 겨울에 건강을 위해 필요한 견과류를 같이 넣고 무친다. 생시금치와 견과류가 주는 고소함과 달달함이 젓갈의 감칠맛과 만나 내는 그 맛에 반한다. 겨울 추위에 감사의 인사를 전하고 싶은 맛이다.

시금치는 예쁜 초록색의 잎이 쭉쭉 뻗어 고운 것보다 붉은색이 도는 거친 잎을 가진 것이 더 달고 고소하다. 그러나 예쁜 것만 좋아하는 사람들에 의해 점차 사라지고 있어 아쉬운 마음이 든다.

In winter, it is very difficult to find green vegetables in the fields. Spinach is almost the only vegetable we can find in such a season. Spinach is a cold-resistant crop, surviving in open fields covered with white snow without freezing to death. The colder it gets, the lower its posture, lying flat on the ground without sense of volume as if it were a picture. When we scrape the spinach out from the frozen ground with a knife and make it into *namul* (seasoned vegetables), it tastes like sweet juice is flowing. Use up to the pink roots after trimming to make fresh kimchi (*geotjeori*). Mix with nuts needed for one's health in the dry winter. You will fall in love with the combination of the nutty, sweet taste of fresh spinach and nuts and the savory taste from the salted fish sauce. It is a taste that makes you thankful for the

winter coldness.

Spinach with rough, reddish leaves is sweeter and more nutty than spinach with beautiful green and straight leaves. However, it is a pity that this delicious spinach is gradually disappearing because people only like pretty things.

재료

손질한 시금치 200g, 통깨 1큰술, 다진 견과류 2큰술

겉절이 양념: 멸치 액젓 1큰술, 고춧가루 1큰술, 다진 파 1큰술, 다진 마늘 1작은술, 생강술 1작은술

만드는 법

1 시금치를 다듬어 씻는다.
2 씻은 후 물기를 뺀 시금치를 먹기 좋은 크기로 자른다.
3 그릇에 겉절이 양념 재료를 모두 넣고 고루 섞는다.
4 시금치에 겉절이 양념을 넣고 잘 무친다.
5 통깨와 다진 견과류를 넣고 다시 한번 무친다.

Ingredients

Trimmed spinach 200g, roasted sesame seeds 1 tablespoon, chopped nuts 2 tablespoons

Fresh kimchi seasoning: salted anchovy sauce 1 tablespoon, red pepper powder 1 tablespoon, chopped spring onion 1 tablespoon, minced garlic 1 teaspoon, ginger liquor 1 teaspoon

Instructions

1 Trim and wash the spinach.
2 Cut the washed and drained spinach into bite-sized pieces.
3 In a bowl, mix all the fresh kimchi seasoning ingredients evenly.
4 Add the fresh kimchi seasoning to the spinach and mix well.
5 Add the sesame seeds and chopped nuts, and mix again.

42

동치미

Dongchimi
White Radish Kimchi in Water

동치미는 김장용 무가 좀 덜 자라 주먹만 할 때 캐서 담그거나, 김장용 무 중 작은 것을 골라 담근다. 통째로 담그기 때문에 아무리 소금에 굴려 절였다가 담가도 간이 무의 속까지 배고 맛이 들기까지 꽤 오랜 시간이 걸린다. 그러니 동치미가 익기 전에 상하는 일이 없도록 염도가 좀 높은 소금물에 담갔다가 물로 희석해 간을 맞춰 먹는다. 대략 동지 무렵에는 맛이 들어야 하므로 11월 초나 중순, 김장을 하기 전에 담그게 된다. 동치미가 익으면 무의 색이 약간 누렇게 변하고 국물에서는 시큼한 향이 기포와 함께 올라온다. 무 한 개를 꺼내 와 썰고 깊이가 있는 볼에 담은 뒤 같이 떠 온 국물의 간을 맹물로 맞춰 부어 먹으면 톡 쏘는 청량함이 배 속까지 느껴진다.

동치미 국물에 국수를 말아 먹어도 좋고 죽이나 누룽지를 끓여 함께 먹으면 더 좋다. 그래서 해마다 동지팥죽을 쑤어 동치미와 함께 먹는다. 넉넉히 끓인 팥죽을 이웃들과 나누면 팥죽보다 동치미 맛있다는 이야기를 더 많이 듣는다.

동치미가 익기까지 오랜 시간을 기다리기가 힘든 사람은 무를 손가락 크기로 썰어서 담그면 일주일 정도만 숙성시켜도 먹을 수 있다. 냉면집이나 여타의 식당에서는 썰어서 담그는 방식을 선호한다.

White radish kimchi in water (*dongchimi*) is made with radishes for *gimjang* that are slightly less mature, about the size of a fist, or by selecting smaller ones from *gimjang* radishes. Since whole radishes are used for making *dongchimi* even though rolled in salt, it takes a considerable amount of time for the saltiness to permeate the white radishes and for the flavor to develop. Therefore, to prevent *dongchimi* from spoiling before it fully ripens, it is made with a

slightly higher salt concentration and diluted with water to taste when served. *Dongchimi* is typically made in early to mid-November before *gimjang*, as it should reach its desired taste around the winter solstice. When *dongchimi* ferments, the radishes turn slightly yellowish, and a sour aroma rises from the *dongchimi* liquid along with the bubbles. To serve, take out one white radish, slice it, place it in a deep bowl, adjust the salinity of the kimchi liquid by adding plain water, and enjoy the refreshing coolness down to your stomach.

It is also excellent with noodles and pairs exceptionally well with porridge or *nurungji* (scorched rice). That's why I make *dongji patjuk* (red bean porridge) every year and have it with *dongchimi*. Whenever I generously share my cooked red bean porridge with neighbors, I receive more compliments about the *dongchimi* than the red bean porridge.

If you find it difficult to wait for *dongchimi* made with the white radishes, you can make it by cutting the radishes into finger-sized pieces, which will ripen in about a week for eating. *Naengmyeon* (cold noodle) restaurants and other restaurants often prefer to make *dongchimi* by cutting the white radishes into pieces.

재료

동치미 무 4개, 소금 100g, 쪽파 40g, 갓 40g, 불린 청각 40g, 배(또는 사과) 1개, 마늘 8쪽, 생강 1쪽, 삭힌 고추 4개, 마른 고추 2개

동치미 국물: 물 8L, 소금 150g

만드는 법

1   무는 주먹만 하게 작고 단단한 것으로 골라 잔뿌리를 제거하고 깨끗하게 씻는다.
    ① 무를 손가락 길이로 썬다.
    ② 썬 무에 소금 100g을 넣어 100분간 절인다.
    ③ 절인 무를 맑은 물에 씻어 건진다.
2   쪽파와 갓은 통으로 깨끗하게 다듬어 씻어 무와 같은 길이로 썬다.
3   불린 청각은 깨끗이 씻어 건져 물기를 제거한다.
4   삭힌 고추와 마른 고추는 물에 씻는다.
5   마늘과 생강은 깨끗이 씻어 편으로 썬다.
6   배(또는 사과)는 씻어 껍질째 4등분한다. (사과와 배를 둘 다 넣어도 좋다.)
7   천으로 만든 주머니에 배, 마늘, 생강, 청각을 넣어 묶는다.
8   물과 소금을 섞어 동치미 국물을 만든다.
9   김치통에 절인 무와 쪽파, 삭힌 고추, 마른 고추, 갓을 담는다.
10 7의 양념 주머니를 넣고 8의 동치미 국물을 붓는다.
11 상온에서 80% 정도 익힌 후 냉장고에 넣어두고 먹는다.

Ingredients

Dongchimi radishes 4, salt 100g, green onions 40g, mustard leaves 40g, macerated sea staghorn 40g, pear (or apple) 1, garlic 8 cloves, ginger 1 piece, fermented pepper 4, dried red pepper 2

Dongchimi liquid: water 8L, salt 150g

Instructions

1   Choose small, fist-sized, firm white radishes for *dongchimi*. Remove the rootlets and wash them.
    ① Cut the white radishes into finger-length pieces.
    ② Sprinkle 100g of salt on the sliced white radishes and salt for 100 minutes.
    ③ Rinse the salted white radishes in clear water and drain.
2   Trim and wash the green onions and mustard leaves, then cut them into lengths similar to the radish.
3   Wash the sea staghorn and drain.

4   Wash the fermented peppers and dried red peppers.

5   Wash and slice the garlic and ginger.

6   Wash the pear (or apple), cut them into 4 pieces with the skin on.

7   Put the pear (or apple), garlic, ginger and sea staghorn into the cotton cloth bag and tie it up.

8   Mix the water and salt to make *dongchimi* liquid.

9   Place the salted white radishes, green onions, fermented peppers, dried red peppers, and mustard leaves in a kimchi container.

10 Put the cotton cloth bag from Step 7 in the kimchi container and pour the *dongchimi* liquid from Step 8.

11 Allow the *dongchimi* to ripen about 80% at room temperature, then store it in the refrigerator and enjoy.

43

무생채

밖에서 밥을 사 먹으면 자주 만나는 무생채는 너무 달고 시다. 맛있는 무 맛에 먹는 것이 아니라 달고 신 양념 맛에 먹는 것 같아 나는 별로다. 무생채도 겉절이로 접근해 단맛 없이 버무리면 밥반찬으로 꽤 괜찮다. 달지 않게 버무려서 마지막에 식초를 조금 넣어 신맛은 내지 않고 상큼함만 더한다. 그 작은 양이 감칠맛도 올린다.

Julienned white radish fresh kimchi (*musaengchae*) served in restaurants is too sweet and sour for my taste. It feels like eating the sweet and sour seasoning rather than enjoying the delicious radish. If you make *musaengchae* as a type of fresh kimchi and season it without adding sweetness, it becomes an excellent side dish. Tossing the *musaengchae* with unsweetened seasoning and adding vinegar at the end adds freshness without making it too sour. This small amount of vinegar also enhances the savory flavor.

재료

무 500g

무침 양념: 고춧가루 2큰술,
멸치 액젓 2큰술, 송송 썬
대파 1큰술, 다진 마늘
1작은술, 식초 1작은술,
깨소금 1큰술

Ingredients

White radish 500g

Seasoning: red pepper
powder 2 tablespoons,
salted anchovy sauce
2 tablespoons, finely
chopped large spring
onion 1 tablespoon,
minced garlic 1 teaspoon,
vinegar 1 teaspoon,
ground sesame seeds 1
tablespoon

만드는 법

1   무는 깨끗이 씻어 채 썬다.

2   무침 양념을 만든다.

3   무침 양념에 준비한 무채를 버무려 낸다.

Instructions

1   Wash and julienne the white radish.

2   Prepare the seasoning.

3   Mix the julienned white radish with the seasoning.

44

배무생채

Bae Musaengchae

Pear-White Radish Fresh Kimchi

단맛을 좋아하는 현대인에게 감미료 없이 달게 만들어 먹는 무생채가 바로 배무생채다. 배와 무를 동량으로 채 썰어 무치면 아삭하고, 재료만으로도 단맛이 풍부한 무생채가 되어 아이들도 잘 먹는다.

For modern people who love sweetness, pear-white radish fresh kimchi (*bae musaengchae*) is a sweet and refreshing dish without any added sweeteners. Julienne equal amounts of pear and white radish and mix them together to create a crunchy and naturally rich sweet *musaengchae* that even children will enjoy.

재료

배 250g, 무 250g, 통깨
1큰술, 식초 1작은술

무침 양념: 고춧가루 2큰술,
멸치 액젓 2큰술, 다진 대파
1큰술, 다진 마늘 1작은술

만드는 법

1 배는 껍질을 벗겨 채 썬다.
2 무는 깨끗이 씻어 채 썬다.
3 그릇에 무침 양념 재료를 모두 넣고 섞는다.
4 양념에 채 썬 무를 넣어 무친다.
5 채 썬 배를 넣고 다시 한번 살살 버무린다.
6 식초와 통깨를 넣어 마무리한다.

Ingredients

Pear 250g, white radish
250g, roasted sesame
seeds 1 tablespoon,
vinegar 1 teaspoon

Seasoning: red pepper
powder 2 tablespoons,
salted anchovy sauce 2
tablespoons, chopped
large spring onion 1
tablespoon, minced garlic
1 teaspoon

Instructions

1 Peel the pear and julienne it.
2 Wash the white radish and julienne it.
3 Place all the seasoning ingredients in a bowl and mix together.
4 Add the julienned radish to the seasoning and mix well.
5 Add the julienned pear and mix gently once more.
6 Finish by adding vinegar and sesame seeds.

45

파래김치

Parae Kimchi

Grassy Seaweed Kimchi

시어머니께 배운 유일한 김치다. 처음엔 이상했지만 매번 더 맛있게 먹는 김치가 되었다. 파래를 씻는 일이 좀 번거롭지만 마음먹고 좀 넉넉히 준비해서 냉장고에 넣어두면 별미를 먹고 싶을 때 찾게 된다. 묵은지 국물이 식재료가 되어 새로운 김치로 탄생한 창의적인 김치가 바로 파래김치다. 처음엔 파래무침 같다가 익으면서 김치 국물과 하나로 어우러지면 세상에 다시없는 새로운 맛의 세계로 넘어간다.

This is the only kimchi recipe I learned from my mother-in-law. At first, it was strange, but it has become a kimchi that I enjoy more each time. Washing grassy seaweed can be a bit cumbersome, but if you prepare a good amount and store it in the refrigerator, you will find yourself reaching for it whenever you crave a unique taste. Grassy seaweed kimchi is a creative dish where _gimjang_ kimchi transforms into an ingredient for a new type of kimchi. It starts off tasting like _parae muchim_ (seasoned grassy seaweed), but as it ferments and blends with the kimchi liquid, it transforms into a whole new world of flavor.

재료

묵은지 국물 2컵, 파래 200g,
대파 1/2뿌리, 쪽파 3뿌리,
마늘 1쪽, 고춧가루 약간,
액젓 약간, 통깨(선택)

만드는 법

1 묵은지의 남은 국물을 버리지 말고 모아둔다.

2 파래는 모래가 나오지 않을 때까지 깨끗하게 씻어 한두 번 자른다.

3 대파는 다듬어 씻어 송송 썰고, 마늘은 곱게 찧는다.

4 쪽파는 다듬어 씻은 다음 2cm 길이로 썬다.

5 묵은지 국물에 파래를 넣고 잘 섞는다.

6 파래에 대파, 쪽파, 마늘을 넣고 잘 버무린다.

7 모자라는 색은 고춧가루를 넣어 내고, 모자라는 간은 액젓으로 한다.

8 먹을 때 통깨를 뿌려 내면 고소하다.

Ingredients

Aged kimchi liquid 2
cups, grassy seaweed
200g, large spring
onion 1/2, green onion
3, garlic 1 clove, red
pepper powder (a small
amount), salted fish
sauce (a small amount),
roasted sesame seeds
(optional)

Instructions

1 Save the leftover kimchi liquid from the aged kimchi.

2 Wash the grassy seaweed thoroughly until there is no sand left, then cut it once or twice into manageable pieces.

3 Trim, wash, and finely chop the large spring onion. Mince the garlic.

4 Trim and wash the green onions, then cut them into 2cm lengths.

5 Add the grassy seaweed to the aged kimchi liquid and mix well.

6 Add the large spring onion, green onions, and garlic to the grassy seaweed and mix well.

7 If needed, add red pepper powder for color and salted fish sauce for taste.

8 For a nutty flavor, sprinkle with sesame seeds before serving.

파래를 씻는 일이 좀 번거롭지만 마음먹고 좀 넉넉히 준비해서 냉장고에 넣어두면 별미를 먹고 싶을 때 찾게 된다.
Washing grassy seaweed can be a bit cumbersome, but if you prepare a good amount and store it in the refrigerator, you will find yourself reaching for it whenever you crave a unique taste.

김치 책

셰프들의 김치 선생님,
고은정의 기본 김치 레시피

초판 1쇄 발행 2026년 3월 3일
초판 2쇄 발행 2026년 4월 17일

지은이 고은정
펴낸이 안지선

사진 류관희
영문 번역 윤금진
영어 감수 토드 캐머런 태커
편집 신정진
디자인 석윤이
마케팅 타인의취향 김경민, 강지민, 강민지
경영지원 강미연

펴낸곳 (주)몽스북
출판등록 2018년 10월 22일 제2018-000212호
주소 서울시 강남구 테헤란로 151, 1006호
이메일 monsbook33@gmail.com

mons (주)몽스북은 생활 철학, 미식, 환경,
디자인, 리빙 등 일상의 의미와 라이프스타일의
가치를 담은 창작물을 소개합니다.

The Kimchi Book

Mastering The Basics of Kimchi with
The Chefs' Teacher

First Edition Printed on March 3, 2026
Second Edition Printed on April 17, 2026
Author Goh Eun-jeong
Publisher Ahn Ji-seon
Photography Ryu Gwan-hee
English Translation Yoon Geum-jin
English Review Todd Cameron Thacker
Editor Shin Jeong-jin
Design Suk Yoony
Marketing Kim Kyung-min, Kang Ji-min, Kang Min-ji
Business Administration Kang Mi-yeon

Published by Monsbook Co., Ltd.
Publication Registration No. 2018-000212 (October 22, 2018)
Address Suite 1006, 151 Teheran-ro, Gangnam-gu, Seoul, Republic of Korea
Email monsbook33@gmail.com

Monsbook Co., Ltd. publishes creative works that explore the meaning of everyday life and the value of lifestyle through themes such as life philosophy, gastronomy, environment, design, and living.